T0262544

HERMAPHRODITISM

HERMAPHRODITISM

*A Primer on the Biology, Ecology,
and Evolution of Dual Sexuality*

JOHN C. AVISE

WITH ORGANISMAL LINE-DRAWINGS
BY TRUDY NICHOLSON

COLUMBIA
UNIVERSITY PRESS
NEW YORK

Columbia University Press
Publishers Since 1893
New York   Chichester, West Sussex
Copyright © 2011 Columbia University Press
All rights reserved

Library of Congress Cataloging-in-Publication Data
Avise, John C.
    Hermaphroditism : a primer on the biology, ecology, and evolution of dual sexuality /
John C. Avise.
            p.   cm.
    Includes bibliographical references and index.
    ISBN 978-0-231-15386-7 (cloth : alk, paper) — ISBN 978-0-231-52715-6 (ebook)
    1. Intersexuality.   2. Intersexuality in animals.   I. Title.
    QP267.A95   2011
    616.6'94—dc22   2010039958

∞
Columbia University Press books are printed on permanent and durable acid-free paper.
This book is printed on paper with recycled content.
Printed in the United States of America

c 10 9 8 7 6 5 4 3 2 1

References to Internet Web sites (URLs) were accurate at the time of writing. Neither
the author nor Columbia University Press is responsible for URLs that may have
expired or changed since the manuscript was prepared.

Cover images: Front, Clark's anemonefish (*Amphiprion clarkii*); *back,* brown garden snail (*Cantareus aspersus*) and Oregon checkerbloom (*Sidalcea oregana*). In these and many other hermaphroditic plant and animal species, each dual-sex individual can reproduce both as a male and as a female during its lifetime.

To J.J. and Buddy,
who remind me every day
to retain a sense of self-worth
and find pleasure in simple things

CONTENTS

PREFACE

The Phenomenon of Dual Sexuality

Who among us has not fantasized, at one time or another, about becoming a member of the opposite sex? If you are a man, what would it feel like to be a woman, if only for a day, a week, or a year? And if you are a woman, what would it feel like to be a man? For most people, such fantasies are idle because our genes and hormones normally dictate that each of us remains of one sex from conception to death. However, for individuals in several hundred species of fish, as well as for those that comprise numerous species of plants and invertebrate animals, experiencing life both as a male and as a female is the reproductive norm. Such dual-sex organisms are hermaphrodites.

Among the vertebrates (animals with backbones), approximately 99% of all species consist of separate-sex individuals, meaning that each individual is either male or female. Most of the other 1% of vertebrate species are hermaphroditic,[1] and essentially all of these are fishes. In some hermaphroditic fish species, an individual may begin its life as a functional male and then later switch to become a functional female, whereas in other species most individuals begin life as females before perhaps later transforming into functional males. In a few fish species, particular individuals switch back and forth repeatedly between male and female. And in still other fish species,

---

[1]A few species among the "other 1%" consist solely of females who reproduce asexually, or clonally (i.e., by parthenogenesis or related reproductive modes). These all-female species are the subject of a companion book to which interested readers are referred: *Clonality: The Genetics, Ecology, and Evolution of Sexual Abstinence in Vertebrate Animals* (Avise 2008).

an individual can be both a functional male and a female at the same time. In some of these simultaneous hermaphrodites, an individual may alternate between male and female behavioral roles during a spawning episode with another hermaphrodite. Finally, in one small taxonomic group of fishes, each hermaphroditic individual normally reproduces by fertilizing itself. When such "selfing"—an intensely incestuous behavior—is repeated generation after generation, the result is a highly inbred lineage, the members of which can become so genetically uniform as to be, in effect, clonally identical to one another.

All of these and additional hermaphroditic phenomena in fishes find their near-perfect analogues in many species of plants and invertebrate animals that also display various forms of dual sexuality. For example, approximately 95% of all species of flowering plants (angiosperms) include at least some dual-sex individuals (hermaphrodites), as do more than 50,000 species of invertebrates. In general, the reproductive lifestyles of hermaphroditic organisms can seem outlandish to us humans, who are more accustomed to thinking of the two sexes being housed in separate bodies.

Hermaphroditism in Fiction

Dual sexuality apparently held a special fascination for the ancient Greeks, as gauged by its prominence in their classical mythologies. Indeed, the word *hermaphrodite* derives from the Greek myth of Hermaphroditus, a son of Hermes and Aphrodite, the ancients' gods of male and female sexuality. As a boy, Hermaphroditus was raised by nymphs on sacred Mount Ida, in what now is the country of Turkey. At age 15, while swimming naked in a pool near Halicarnassus (modern Bodrum, Turkey), the handsome young man was admired and soon accosted by a lovely naiad—Salmacis—who embraced him, kissed him, and prayed to the gods that they be united forever. Her wish was granted; their bodies blended and Hermaphroditus thereafter was part male and part female, simultaneously. In Greek paintings, Hermaphroditus is often depicted as a winged youth with mixtures of male attributes (including male genitalia) and female attributes (including hairstyle, breasts, and broad thighs).

Tiresias—the blind prophet-priest of Zeus—was another type of hermaphrodite who in this case switched sequentially between male and female. It all began when Tiresias chanced upon a pair of copulating snakes and beat them with a stick. Hera (the wife of Zeus) was infuriated, and punished Tiresias by transforming him into a woman. Tiresias then married, had children, and according to some accounts became a female prostitute of considerable renown. Seven years later, Tiresias again encountered two mating snakes, but this time she left the serpents alone. As a reward, Hera permitted Tiresias to regain the male condition. In an interesting footnote to this story, Tiresias became embroiled one day in an argument between Hera and Zeus about who enjoyed sex more: men (as Hera claimed), or women (as Zeus claimed). Having experienced sex both ways, Tiresias was posed the question, and he answered that the female receives ten times more pleasure. Infuriated by this response, Hera struck Tiresias blind. Zeus could not stop this action, but, in partial compensation, he was able to grant Tiresias the gift of foresight, thereby enabling Tiresias to become a seer.

Agdistis was another hermaphrodite in Greek mythology. (S)he was conceived one day when Zeus fell asleep upon the ground and some of his leaked seed (semen) accidentally impregnated Gaia (mother Earth). Agdistis was a superhuman being, simultaneously male and female. However, the gods feared Agdistis and soon castrated it.

In another story from ancient Greece, hermaphroditism was humanity's original state. However, Zeus then decided to split the hermaphrodites, only to find that the now-separate genders spent most of their time in futile efforts to reunite. So, Zeus decided to reconstruct the genitalia such that when men and women embraced, they would fit nicely together and thereby conceive and propagate. And so it has been ever since.

Hermaphroditic or androgynous beings are similarly embedded in the religious traditions or mythologies of several other cultures around the world. For example, in one interpretation of Genesis in the Bible, Adam's body was hermaphroditic originally, but later it was cleaved or partitioned into male and female (Eve) as a part of the fall from grace in the Garden of Eden. In the tribal stories of some native North Americans, a mythological Trickster is mostly male but dresses as a female and gives birth to children. He carries his penis in a box, which he sends to women for purposes of intercourse. In the Dogon peoples of the Mali region in West Africa, a newborn baby who touches male and female outlines—drawn in the sand by a mythical figure—becomes possessed by two souls and may remain androgynous, without a strong proclivity to procreate.

India is especially rich in mythologies entailing dual sexuality. The great Hindu deity Shiva is often portrayed as partially fused with his female alter ego, Parvati. In Buddhism, a male Bodhisattva (a person who has attained Enlightenment) named Avalokitesvara later became a female, Guan Yin. There are East Indian legends in which ancient hermaphrodites are replaced by twins; of individuals who switch month to month between king and queen; of men who were made woman-like by a god's curse; and of males bearing children. A central feature of Tantrism is the desirability of activating both the male and female components of a person's inner self, thereby bringing greater wholeness to the life experience. Sequential hermaphroditism can even be intergenerational. For example, a central idea in Hinduism is that souls transmigrate such that an individual is reborn time and again, often in opposite sex to that of the previous life.

Hermaphroditism also recurs as a theme in modern fiction. In *Star Wars*, Hutts are hermaphrodites, as are Hermats in the series *Star Trek: New Frontier*. In the novel *Raptor* by Gary Jennings (1993), the main character is a hermaphrodite, and so too is the protagonist in the book *Middlesex* by Calliope Stephanides (2002). In the novel *The Left Hand of Darkness* by Ursula Le Guin (1966), a planet is populated by sequential androgynes. For 24 days each month, these people are sexually inactive, but for the ensuing two days they become either male or female as determined by appropriate negotiation with an interested sex partner.

Hermaphroditism in the Real World

As intriguing as such mythological dual-sex characters may be, they hardly outmatch the marvelous hermaphroditic plants and animals that actually do populate our planet, and that are the subject of this book.

In this book I provide an introductory overview of real-life hermaphroditic organisms. I pay special attention to ecological and evolutionary explanations for the remarkable reproductive behaviors and lifestyles of these dual-sex creatures, which in some respects seem to have achieved a best-of-two-worlds outcome by combining male and female functions within each individual. The book will convey two central themes: hermaphroditic species can be highly successful ecologically and evolutionarily; and their successes (and failures) offer fresh perspectives on the adaptive significance of alternative sexual systems.

Beyond conveying these basic biological messages, this book is meant to be entertaining as well as educational for a wide audience. It is intended for college students, teachers, natural historians, and others who would appreciate a nontechnical introduction to the biology of dual-sex organisms. Hermaphroditism has been addressed in many previous works, but typically either from a mostly theoretical perspective (e.g., Charnov 1982) or as a small and specialized component of much broader topics such as sexual allocation, sperm competition, or sexual selection (Birkhead and Møller 1998). To my knowledge, no other book occupies the current niche: a biology-rich overview of the natural history, ecology, and evolution of the phenomenon of dual sexuality.

Chapter 1 sets the general biological stage by asking what defines maleness versus femaleness in diverse organisms. It contrasts hermaphroditism with the standard separate-sex condition (termed *gonochorism* in animals and *dioecy* in plants), and outlines the many ecological and evolutionary topics for which the issue of hermaphroditism is highly germane. The remainder of the book canvases the vast scope of hermaphroditism in the biological world, beginning with plants (chapter 2) and invertebrate animals (chapter 3), and then building on these observations and concepts to describe various hermaphroditic phenomena in fishes (chapter 4). All of the chapters contribute to the broader thesis that hermaphroditism is not only fascinating in its own right but also provides a unique biological vantage for reexamining the ecological and evolutionary significance of familiar separate-sex reproduction.

Trudy Nicholson provided the line drawings of animals and plants that grace this book (figs. 2.4–2.9, 3.1, 3.2, 3.4–3.9, 3.12, 3.14–3.16, 4.2–4.9, 4.16–4.22, and the frontispiece). She has illustrated several of my books on natural history, and it is always a joy for me to work with this gifted and conscientious artist. Judith Mank kindly provided unpublished information, from her earlier dissertation research, on the taxonomic distribution of hermaphroditism in fishes. Thanks also go to Joan Avise for several forms of assistance, and to Felipe Barreto, Rosemary Byrne, Diane Campbell, Jinxian Liu, Ann Sakai, Andrei Tatarenkov, Steve Weller, and several anonymous reviewers for helpful comments on various drafts of the manuscript.

HERMAPHRODITISM

CHAPTER ONE

Two Sexes in One

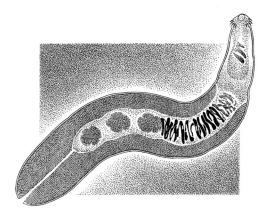

A hermaphrodite is an individual that produces functional male gametes *and* female gametes (sex cells) during its lifetime. This capacity to reproduce both as male and female has many biological ramifications. It raises, for example, the issue of what defines a male versus a female in any species, which in turn motivates the question of what constitutes a male gamete versus a female gamete. With respect to ontogeny (individual development), hermaphroditism begs questions about the genetic and environmental determinants as well as the proximate hormonal and physiological underpinnings of dual sexuality. With respect to ethology (behavior), hermaphroditism raises questions about sexual roles, such as what reproductive tactics an individual might employ to enhance its genetic fitness by investing in reproduction as a male versus as a female at various times during its life. With respect to genetics, hermaphroditism raises questions about the consequences of self-fertilization versus outcrossing for each dual-sex specimen's fitness, and for the broader genetic architectures of populations of dual-sex individuals. With respect to natural history, hermaphroditism raises many questions about how natural selection views different modes of reproduction by dual-sex organisms. With respect to ecology, hermaphroditism touches on topics related to many sexual phenomena, such as population sex ratios and the operation of sexual selection (a form of selection arising from differences in mating success among conspecific individuals). Finally, with regard to evolutionary history, hermaphroditism raises questions about the phylogenetic origins and transformations among alternative reproductive modes, and about the selective forces that sometimes appear to have favored a long-term retention of the dual-sex lifestyle. This book addresses all of these and related topics.

All forms of hermaphroditism entail the joint production by an individual of male and female gametes. Depending on the species or situation, the gametes that unite to form each zygote (fertilized egg) might come from the same hermaphroditic individual (via self-fertilization) or from separate individuals (via outcrossing). In either case, hermaphroditic species are sexual reproducers, as opposed to asexual or clonal reproducers. Thus, they join the vast majority (99.9%) of multicellular animal and plant species that engage routinely in sex, i.e., in cellular processes that entail the regular production (during gametogenesis) and union (during fertilization or syngamy) of male and female gametes (Williams 1975). From an evolutionary-genetic

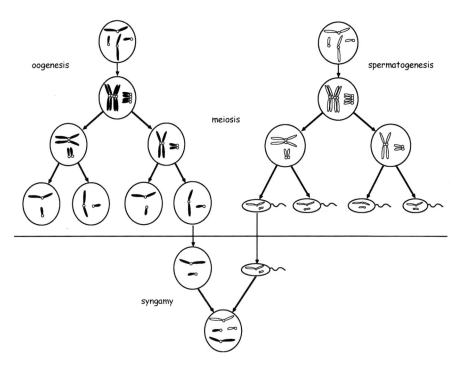

FIGURE 1.1   Simplified diagram of meiosis (during gametogenesis) and syngamy (fertilization), which characterize sexual reproduction. In separate-sex animal species, oogenesis (the production of oocytes or unfertilized eggs) and spermatogenesis (the production of sperm) take place in separate female and male individuals. In dual-sex species, each hermaphroditic individual normally has the capacity to produce both male and female gametes, either simultaneously or sequentially.

perspective, sex in multicellular organisms is basically synonymous with the temporal intercalation of meiosis that produces haploid gametes, and syngamy that restores the diploid condition (fig. 1.1). Sexual reproduction greatly facilitates genetic recombination, and clearly has been an extremely successful evolutionary strategy.

Among the vertebrates, true functional hermaphroditism is essentially confined to several hundred species of fishes. However, in both the scientific and popular literature, the word *hermaphrodite* sometimes is encountered in reference to particular human individuals or to occasional atypical specimens in other vertebrate taxa such as fish or frogs. Such cases usually refer to a sexual ambiguity that arises because an individual's genitalia or other sexual features show intermediate conditions (or various combinations) of what normally we deem to distinguish male from female. For example, in some people the phallus is midway in size and shape between a penis and a clitoris; and in other people the labia are partially fused into a scrotum-like structure. Individuals with such intermediate conditions are best termed intersexual people, rather than hermaphrodites (box 1.1). Intersexualities of this general type also occur sporadically in many separate-sex vertebrate species (Armstrong and Marshall 1964). In this book, however, I will restrict the term *hermaphrodite* to plant and animal species in which many or most dual-sex individuals produce functional male and female gametes at one time or another during their lifetimes, and thus in which operational hermaphroditism is a standard mode of reproduction.

BOX 1.1   Intersexual People

In the modern medical literature, *intersex* refers to anomalies of the reproductive system (including internal reproductive organs, external genitals, or genetic and hormonal systems) that make a person look "different" from most males and females. The condition may affect about 1 in 2,000 children (0.05%), but the estimate is provisional because secrecy tends to surround the condition and because what is deemed to qualify as anomalous can be rather subjective. Intersexuality may arise in at least some cases from unusual balances of sex hormones, which may have either genetic or environmental etiologies, or both. Intersexual people formerly were referred to as hermaphrodites or pseudohermaphrodites, but this practice is now discouraged because of social prejudices or sexual biases that such words might convey, and also because most intersex people are

(continued)

BOX 1.1   (*continued*)

---

not true hermaphrodites (under the functional reproductive definition employed in this book).

Fausto-Sterling (2004) relates the story of one interesting intersexual person. In 1843, Levi Suydam was a 23-year-old who petitioned the town board of Salisbury, Connecticut, to vote in a local election. This request was controversial because Levi was viewed as a woman by most of the townspeople, and this was long before the social movement for woman's suffrage. To settle the dispute, a physician was called to examine Levi, who proved to have a substantial phallus and thus was declared male. The physician's diagnosis was premature, however, as he later learned that Levi also menstruated regularly and had a vaginal opening. Levi also had narrow shoulders, broad hips, and many mental dispositions that the good physician deemed characteristically feminine. Young (1937) relates another such interesting story. Emma had grown up as a female, but had both a penis-size clitoris and a vagina. As a teenager, Emma was attracted to and had sex with several girls, but at age 19 (s)he married a man, with whom (s)he also had pleasurable sex.

Intersexual people may be "straight," "gay," or "bisexual." They sometimes find themselves stigmatized by society and perhaps even by the medical profession itself; intersexual patients are routinely subjected to surgical procedures (typically on children) to reconstruct or conceal what are interpreted to be physical anomalies. In recent years, a growing intersex social movement has promoted several recommendations for parents: raise an intersexual child as either a boy or a girl based on the best available information about his or her personal preference; delay all nonemergency surgeries at least until a child is old enough to make an informed decision; and be open to gender adjustments if and when a child decides to live a sex different from the original assignment. For more on human intersexuality and the life experiences of intersex people, see Dreger (1999, 2001), Fausto-Sterling (2000), and Hines (2003).

Maleness and Femaleness

In humans and other mammals, males typically are XY and females are XX with respect to the sex chromosomes. Thus, males are the heterogametic sex (with two different types of sex chromosomes) whereas females are the homogametic sex (with two copies of one type of sex chromosome). Following gametogenesis in diploid females, each haploid oocyte or egg carries one copy of the X-chromosome; and following gametogenesis in diploid males, each haploid sperm carries either an X-chromosome or a Y-chromosome with about equal likelihood (given the segregation rule of Mendelian inheritance). Thus, whether a boy or a girl results from a particular conception is dictated by whether a Y-bearing sperm or an X-bearing sperm fertilizes the focal egg. Remarkably, many dioecious (separate-sex) plant species have chromosomal modes of sex determination that are "strikingly similar" to the familiar XY systems of mammals and some other animals (Charlesworth 2002).

However, the XY and XX conditions cannot provide universal definitions or proxies for maleness and femaleness because non-mammalian species have a diversity of other sex-determination mechanisms (Bull 1983). For example, females are the heterogametic sex and males are the homogametic sex in birds and butterflies. (By convention, females in such taxonomic groups are denoted as ZW and males as ZZ.) In poikilothermic (cold-blooded) vertebrates—most fishes, reptiles, and amphibians—some separate-sex species are male-heterogametic (XY), some are female-heterogametic (ZW), some have no apparent sex chromosomes, and others display temperature-dependent sex determination (TDSD) in which an individual's gender is strongly influenced by its rearing temperature (box 1.2). Various of these genetic and environmental mechanisms of sex determination sometimes

---

BOX 1.2   Temperature-dependent Sex Determination (TDSD)

The Greek philosopher Aristotle thought that the female sex is inherently cold and the male sex is inherently hot. Thus, when cold winds blow, most conceptions are of females, and warm breezes yield mostly male embryos. Today, we know that if Aristotle had been referring to reptilian embryology (he was not), he could have been partially correct. In alligators, for example, which lay clutches of eggs in scooped-out nests of sand, embryos incubated at temperatures of <30°C develop into females, whereas embryos reared at >34°C become males. Intermediate temperatures yield clutches with mixtures of males and females (as separate individuals). However, any suggestion that Aristotle had great foresight about sex determination, even in reptiles, is quickly negated by the nature of TDSD in turtles. In many of these reptiles, which also lay their clutches in sand, low incubation temperatures yield mostly male offspring whereas high incubation temperatures yield mostly females.

TDSD is common in nesting reptiles (Shine 1999) but also occurs sporadically in fishes (e.g., Conover 1984; Desperz and Melard 1998) and

(continued)

BOX 1.2 (*continued*)

other taxa. It provides a striking example of environmental sex determination (ESD), which contrasts with the chromosomally hardwired systems of XY and ZW genetic sex determination (GSD) in mammals and birds (and many other organisms). In truth, however, ESD and GSD are endpoints on a conceptual continuum, with genetic and environmental factors sometimes interacting to determine an individual's sex (Valenzuela et al. 2003; DeWoody et al. 2010). For example, in some fish species with separate sexes, genotypes and rearing temperatures jointly influence the sexual outcomes (Schultz 1993; Patino et al. 1996).

The observation that exogenous factors influence sexual development in numerous separate-sex vertebrates is also of interest in the current context because it will segue (in chapter 4) into the notion that environmental factors influence patterns of sexual expression in many sequentially hermaphroditic fishes as well.

---

co-occur within taxonomic orders, families, genera, and even particular species of fish (Devlin and Nagahama 2002), thus implying a remarkable evolutionary lability in the developmental factors underlying sex (Francis 1992) and in the phylogenetic transformations between alternative sex-determination modes (Mank et al. 2006; Mank and Avise 2009). Much the same can be said for plant species, which collectively display a wide variety of genetic and environmental sex-determination modes. Such evolutionary flexibility in fishes and plants contrasts diametrically with the evolutionary conservatism of sex determination in mammals and birds, almost all species of which show XY and ZW genetic systems, respectively (Foster and Marshall-Graves 1994; Fridolfsson et al. 1998; Lahn and Page 1999). (Exceptions among the mammals include some voles [Just et al. 2007] as well as the platypus and the echidna [Veyrunes et al. 2008].) A general sexual flexibility is also one reason why both piscine and plant lineages (in contrast to avian and mammalian lineages) seem biologically predisposed to the frequent evolution of functional hermaphroditism, as well as to common evolutionary shifts between dual sexuality and the separate-sex condition (chapters 2 and 4).

If the chromosomal or other proximate mode of sex determination is not the deciding criterion for gender in the biological world, what then universally distinguishes a male from a female? Perhaps the gender that produces mobile gametes (e.g., sperm with tails for active swimming, or pollen with a propensity to disperse from the parent plant) is invariably male, and the sex that produces non-mobile gametes (tail-less oocytes or mostly stationary eggs) is invariably female. The mobility criterion works well for vertebrate animals and for many plants, but it can fail elsewhere. In some algae and plants, for example, male and female gametes are both tail-less (unflagellated); whereas in some other such species, both male and female gametes are flagellated and mobile. To distinguish males from females in general, various other possible criteria, such as hormonal profiles, behaviors, physiologies, or morphologies, likewise are far too variable across all plants and animals to suffice as any universal basis for the definition of maleness versus femaleness.

From an evolutionary vantage, the one-and-only phenotypic feature that consistently distinguishes males from females is gamete size. In any multicellular organism, individuals that produce smaller gametes are males, by definition; and individuals that produce the larger gametes are females, by definition. This situation is referred to as anisogamy: the strongly bimodal distribution of gamete size (smaller in males, larger in females) that characterizes the vast majority of sexually reproducing organisms. (In some multicellular algae and fungi, two genetic types of gametes are similar in size, but technically these species do not violate the broader rule because the two genders are referred to as mating types ["+" and "−"] rather than as males and females.)

Anisogamy and the Separate-sex Condition

In multicellular organisms, different types of somatic (body) cells are specialized for different functions. In animals, nerve cells perform very different tasks than blood cells, for example, as do xylem cells versus leaf cells in a plant. All somatic cells within an individual trace back through mitotic cell divisions (a clonal genetic operation) to a single fertilized egg, or zygote. Thus, cellular specializations reflect differences across somatic cell lineages in how particular genes have been regulated (activated, repressed, or otherwise modulated) during an individual's development. Epigenetics refers to any mechanism involved in regulating gene activity during ontogeny, i.e., to any mechanism that causes phenotypic variation without altering the base-pair nucleotide sequences of the genes (Gilbert and Epel 2009).

Cells that play a direct reproductive role are termed sex cells, germ cells, or gametes. More specific labels such as sperm, spermatozoa, or pollen (the male gametophyte) describe the male gametes, and oocytes, ova, or unfertilized eggs describe the female gametes. The haploid cells that engage in syngamy arise from meiotic processes during gametogenesis, which transpires in a germ line that in most animals (as opposed to the situation in most plants) becomes sharply demarcated from somatic cell lines early in embryonic development. Because relative gamete size is the ultimate phenotypic criterion for defining femaleness and maleness, it is interesting to speculate on the evolutionary origins of anisogamy early in the history of multicellular life (Majerus 2003).

Evolutionary Origins

In an evolutionary scenario developed by Parker and colleagues (1972), an ancient arms race among gametes precipitated the transition from isogamy (the presumed ancestral condition of equal-sized gametes) to anisogamy. These researchers envisioned an ancestral population of isogamous organisms in which a *de novo* mutation initially led one individual to produce smaller gametes than the norm. Because its sex cells were smaller, that organism could produce more gametes with the same amount of energetic investment. Any such individual would have enjoyed an initial reproductive (fertilization) advantage over its compatriots, all of whom still produced larger gametes. The small-gamete mutation therefore would increase in frequency in the population, for a time. However, as the mutant allele became more common, the likelihood increased that two small gametes would encounter one another and fuse. The resulting zygote would likely have been debilitated, however, because the small gametes that produced it offered few nutrients for the nascent embryo. Thus, natural selection would have favored any tendency for small gametes to fertilize only large ones; and the resulting competition among small-gamete individuals for fertilization success would select for individuals who produced even smaller (and thus even more) gametes. As tiny gametes become more prevalent in the population, selection pressures would escalate on large-gamete individuals to produce larger gametes to compensate for the limited nutrients that the tiny gametes contribute to a zygote. From this disruptive selection regimen favoring both large gametes and small gametes in the population, anisogamy eventually emerged as an evolutionarily stable outcome.

Hurst (1990) offered a different scenario for the evolutionary origin of anisogamy. It is well known that intracellular parasites such as bacteria often inhabit the cellular cytoplasm, and that they can impose strong evolutionary selective pressures on hosts (Burt and Trivers 2006). Hurst's idea is that anisogamy evolved because of an advantage it conferred (relative to isogamy) in reducing the probability of infection by a disease agent during fertilization. Today, small male gametes (e.g., sperm) typically contribute little or no cytoplasm to the fertilized egg, which instead acquires cytoplasm mostly from the oocyte. Because only one parent (the female) contributes cytoplasm to a zygote, anisogamy appreciably diminishes the probability that a zygote acquires a cytoplasmically housed agent of an STD (a sexually transmitted disease) via syngamy. This advantage probably applied early in the evolutionary history of life as well, and it may have contributed to selection pressures favoring anisogamy.

Another hypothesis for the evolution of anisogamy also builds on the observed disparity between male and female gametes in cytoplasmic contributions to the zygote. Mitochondria are organelles that reside by the hundreds or thousands in the cytoplasm of each somatic and germ cell (box 1.3). They carry tiny genomes (mtDNAs) that encode key components (others of which the nuclear genome encodes) of the molecular machinery by which cells produce chemical energy (fig. 1.2). Mitochondria are not cellular parasites, but like cytoplasmic bacteria they normally are transmitted

---

BOX 1.3 Cytoplasmic Genomes

These are relatively small snippets of DNA housed inside organelles within the cytoplasm of eukaryotes (organisms whose cells have a distinct membrane-bound nucleus). The two primary cytoplasmic genomes are mtDNA within the mitochondria of animals and plants, and cpDNA in plant chloroplasts (Avise 2004). MtDNA is usually a closed-circular molecule of length approximately 16–20 kilobase pairs (kbp, i.e., thousands of base pairs) in most animal species and 200–2500 kbp in various plant species. Normally, it is inherited maternally (matrilineally) from one organismal generation to the next. CpDNA is a closed-circular molecule of length approximately 120–217 kbp in various plants. It too is usually transmitted uniparentally—maternally in most species, but paternally in gymnosperms and some other plants. Occasional instances of biparental inheritance for cpDNA also are known (e.g., Metzlaff et al. 1981). Mitochondria are the powerhouses of cells—the sites of oxidative phosphorylation and electron transport that generate and store (in the form of ATP, or adenosine triphosphate) much of a cell's chemical energy (fig. 1.2). Chloroplasts play key roles in the photosynthetic processes that generate a plant's food from carbon dioxide and water.

Under the widely accepted endosymbiont theory (Margulis 1970), these cytoplasmic genomes are interpreted to be the descendents of free-living prokaryotic cells that early in the history of life invaded (or were engulfed by) proto-eukaryotic cells. This eventuated in a tight collaborative relationship in which the merged cells bore the precursors of the genes now housed in the nuclear and cytoplasmic compartments of eukaryotic cells. Many of the genes initially present in the cytoplasmic endosymbionts gradually transferred to the nucleus, so the genes that remain today in mtDNA or cpDNA are merely "skeleton crews" descended from their original bacterial ancestors. Across evolutionary time, the symbiotic relationship became so intimate that today cytoplasmic genomes could not function independently of the eukaryotic cells of which they have become integral parts, nor could eukaryotic organisms survive without their cytoplasmic organelles.

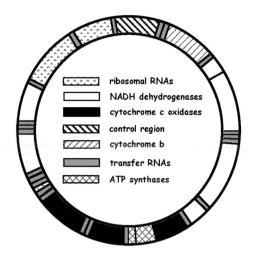

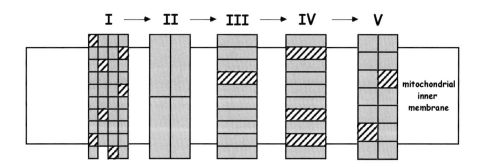

intermembrane space

mitochondrial matrix

FIGURE 1.2 *Above*, the structure of animal mitochondrial DNA. Typically, each mtDNA molecule is composed of 22 loci specifying transfer (t) RNAs, 13 genes encoding protein subunits, and two loci specifying ribosomal (r) RNAs, all of which in effect are maternally transmitted as one linked cytoplasmic "supergene." *Below*, simplified diagram of the molecular machinery involved in producing chemical energy (ATP) along the mitochondrial inner membrane (from Avise 2010, after Graff et al. 1999). For each enzyme complex (I–V), each hatched box represents a distinct polypeptide subunit encoded by mtDNA, and each shaded box indicates a polypeptide encoded by nuclear DNA and then exported to the mitochondrion.

maternally, via the oocyte, from one generation of organisms to the next. This uniparental mode of inheritance led Hurst and Hamilton (1992) to propose an intriguing hypothesis: anisogamy was (and still is) favored by natural selection because it minimizes the potential for intra-zygote conflict between what would otherwise be genetically distinct populations of cytoplasmic organelles delivered by the fusing gametes. Similar kinds of evolutionary arguments in favor of anisogamy can be made with respect to a plant's cytoplasmic organelles, which in addition to mitochondria also include chloroplasts that likewise carry their own unique little genomes (cpDNAs).

These three hypotheses for the evolution of anisogamy are not mutually exclusive, and my own guess is that all of them may have contributed to the evolutionary emergence of the anisogamy phenomenon.

Evolutionary Ramifications

Regardless of exactly how anisogamy evolved, it has had profound consequences for multicellular plants and animals (Parker et al. 1972; Trivers 1972). Most fundamentally, it means that males have an inherent capacity to produce vast numbers of small and energetically cheap gametes, whereas females can produce far fewer and individually more expensive eggs. This typically gives a healthy male the reproductive potential to sire vastly more offspring than an equally healthy female can dam. The degree to which particular males can realize this extra reproductive potential depends on many species-idiosyncratic ecological, behavioral, and evolutionary considerations related to a population's mating system (basically, who mates with whom). The following discussion will pertain especially to many vertebrate species with active mate choice (such as many birds, mammals, and fish); these provided a biological framework for much of the traditional research on sexual selection theory.

Figure 1.3 defines alternative mating systems and summarizes their relevance to how anisogamy ultimately plays out in various separate-sex species in nature. Under polygyny, by definition, particular males may have multiple mates but each female typically has only one mate; under polyandry, particular females have multiple mates but each male has only one; and under polygynandry, members of both genders typically have two to several mates each. (Promiscuity is an extreme form of polygynandry in which each male and female has many mating partners.) These alternative expressions of polygamy are all to be distinguished from monogamy, wherein each male and each female has only one mate. Depending on the context, these definitions can apply within a single mating season or across an individual's reproductive lifetime. A distinction is also to be made between a population's social mating system (involving who mates or "pair-bonds" with

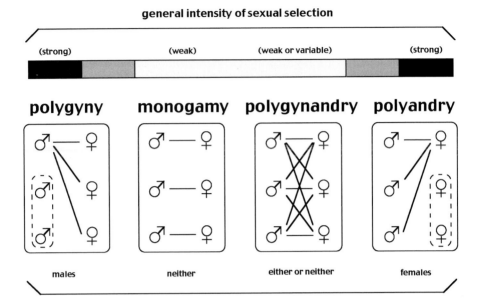

FIGURE 1.3   Schematic definition of four mating systems in separate-sex species. Lines connecting males and females indicate mating partners. Also shown is the theoretical selection gradient in the intensity of sexual selection, and the resulting sexual dimorphism in phenotype that might be expected in secondary sexual traits that have been molded by sexual selection (from Avise et al. 2002). The dashed lines imply that some individuals may not be able to find mates, especially when a mating system entails extreme polygyny or polyandry (see text).

whom) and its genetic mating system (the matings that resulted in success-ful fertilization events and viable offspring) (box 1.4).

In principle, the genetic mating system of a population or species is highly pertinent to the direction and intensity of sexual selection. For ex-ample, in a population with a 1:1 sex ratio and a polygynous mating system, some males inevitably go without mating partners, whereas others may ob-tain several mates, thus leading to pronounced competition for females. This can translate into strong sexual selection on males, and to the evolu-tion of secondary sexual traits (phenotypic traits that arise via sexual selec-tion) in that gender. Conversely, in a population with a 1:1 sex ratio and a polyandrous mating system, some females fail to mate whereas others are multiple maters, thus driving sexual selection on females and perhaps pro-moting the elaboration of secondary sexual traits in that sex. However, the disparity between the genders in the intensity of sexual selection (and in the degree of elaboration of secondary sexual traits) is seldom expected to

BOX 1.4   Molecular Documentation of Biological Parentage

Naturalists traditionally studied animal mating systems by observing the behaviors (sometimes including copulations) of reproductively active males and females. Similarly, the mating systems of plants typically were deduced by monitoring pollinator movements. Such observations revealed the apparent or the social mating system of a population, but not necessarily the realized or the true genetic mating system (which requires knowledge of actual paternity and maternity). The social mating system of a population can be misleading about the genetic mating system if surreptitious reproductive behaviors (such as mate infidelities or other sources of foster parentage) are common, as they are for example in many birds (Griffith et al. 2002), fishes (Avise et al. 2002), and various other taxa (Avise 2004). Similarly in plants, the apparent mating system can be misleading about the true mating system if, for example, the pollen that achieved the fertilizations did not come from the pollinator who was observed to visit the plant. Yet the genetic mating system of any animal or plant species is crucial to understanding the operation and evolution of sexual selection, because it registers the actual intergenerational transmission of genes.

In the mid-1960s, breakthroughs in molecular biology gave scientists the first good opportunities to document genetic parentage (and, thus, genetic mating systems) in nature. Since then, researchers have used molecular markers from various protein-level or DNA-level assays to deduce patterns of genetic paternity and maternity in hundreds of animal and plant populations (Jones and Ardren 2003; Avise 2004). The earliest procedure was protein electrophoresis, followed by several methods of DNA fingerprinting involving, for example, minisatellite and microsatellite loci. The basic approach in each case involves comparing the genotypes of particular progeny with those of candidate parents at each of a series of highly polymorphic loci. From basic rules of Mendelian inheritance, any adult whose genotypes are incompatible with alleles observed in progeny can be genetically excluded as the biological parent. Additional nuances of genetic parentage analysis depend on a variety of statistical and biological considerations, such as the exclusionary power of the molecular markers, the number of candidate parents, and the nature of any independent knowledge about the system (Jones et al. 2009).

The accompanying diagram (from Avise 2002), shows a hypothetical example in which genetic paternity can be partially deduced, even from a single polymorphic locus, for a clutch of offspring with a known mother. The diagram represents a protein-electrophoretic gel or a microsatellite gel from a laboratory assay, and each band is one allele. Each column displays

(continued)

BOX 1.4   (*continued*)

the genotype of one individual, with the leftmost column representing
the known dam of the 19 progeny in lanes to the right. For each offspring,
either the "82" or the "90" allele came from the mother, so the second allele must have come from the biological father. Collectively, there are four
alleles ("86," "88," "92," and "96") of paternal origin in the clutch, so the
brood must have been sired by at least two males (assuming that each
father was heterozygous for a different pair of those four alleles).

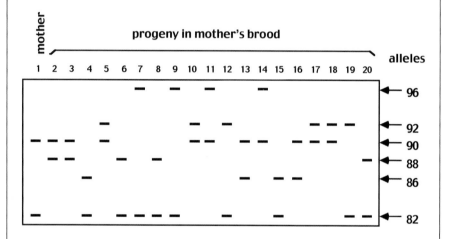

By compiling such information from multiple loci and many families,
reasonable deductions about the genetic mating system of a natural population can often be reached.

be as strongly female-biased under polyandry as it is to be male-biased under polygyny. This is because females inherently remain more fecundity-
limited than males, due ultimately to anisogamy.

Secondary sexual traits are gender-specific phenotypic characters (other
than the primary reproductive organs themselves) that evolved via sexual
selection and tend to be confined to or elaborated mostly in one sex only.
Modern researchers follow Darwin (1871) in categorizing sexual selection as
either intrasexual or intersexual (epigamic). Intrasexual selection (sometimes abbreviated as "male-male competition") typically favors males that
are good fighters, stalwart defenders of territory, or otherwise stiff competitors against other males. It has resulted in such impressive traits as antlers in
bull mooses, tusks in male walruses, and fighting behaviors in gamecocks.

For winning males, the ultimate fitness reward is additional offspring via enhanced mating success. Intersexual or epigamic selection (sometimes abbreviated "female-choice") typically operates by favoring males that females find attractive as mates. It too has resulted in the evolution of impressive phenotypic features, such as peacock tails, showy fins and scales on male guppies, and elaborate male courtship behaviors in many vertebrate and invertebrate species (Andersson 1994). The reproductive payoff, again, is more progeny via enhanced mating access.

Sexual selection can be a powerful evolutionary force promoting striking intraspecific sexual dimorphism (consistent mean differences between conspecific males and females) in secondary sexual traits. Males typically become the more elaborated gender, as in mooses, walruses, peafowls, and guppies. Thus, overall, in many evolutionary lineages of separate-sex organisms, the basic sexual asymmetry in gamete size—anisogamy—ultimately has initiated an evolutionary cascade of biotic consequences that often includes the following:

(a) *higher fecundity potential in males (i.e., a capacity for higher gametic output by individual males than by individual females).*

(b) *behavioral tendencies for males (more so than for females) to seek multiple mates.* By fertilizing the clutches of multiple females, a male potentially stands to increase dramatically the number of offspring he sires, whereas a female who allows her clutch(es) to be fertilized by multiple males cannot normally expect a comparable increase in fitness payoff. The statistical regressions that describe the relationship between mate numbers and progeny production are known as sexual selection gradients or Bateman gradients (box 1.5), which typically are much steeper for males than for females in many animal and plant species. (Notable exceptions to this rule include particular pipefish species and other taxa with sex-role-reversal, in which females may display steeper Bateman gradients than males [Jones et al. 2000].)

(c) *greater tendencies for polygyny than for polyandry to evolve.* Strict polyandry is relatively rare in the biological world, whereas polygyny is common. This makes considerable evolutionary sense, given that steeper Bateman gradients for males than for females characterize many species. Another reason why polyandry is relatively rare has to do with asymmetrical assurances of genetic parentage. Especially in taxa that give birth to live young (such as female-pregnant mammals) or that brood their clutches (such as many nesting birds), a female normally can be quite certain that she is the biological mother of the offspring she nurtures, whereas a male has much less assurance of paternity for the offspring he might help to rear. Because any male that enters into a polyandrous relationship runs the considerable risk of being

BOX 1.5   Bateman Gradients

In the scientific literature on animal mating systems, the relative intensities of sexual selection on males versus females in various populations or species have been attributed to several influences including differences in parental investment (e.g., Trivers 1972; Parker and Simmons 1996), differences in operational sex ratio (the relative available numbers of sexually mature males and females; Kvarnemo and Ahnesjö 1996), relative variances in reproductive success of males and females (Payne 1979; Wade and Arnold 1980; Wade and Shuster 2004), and the potential reproductive rates of the two sexes (Clutton-Brock and Vincent 1991; Clutton-Brock and Parker 1992). In a classic paper that appeared in 1948, Angus Bateman argued that all such influences can be subsumed under one first-order factor or common denominator: the average relationship between the number of mates an individual obtains (its mating success) and the number of offspring it produces (its reproductive success or genetic fitness).

In experimental populations of *Drosophila*, Bateman documented that males' mean genetic fitness increased rapidly as a function of the number of mates acquired, yielding a steep sexual selection gradient; whereas females' genetic fitness did not increase nearly as much with mating success, yielding a nearly flat selection gradient (see the accompanying graph). Bateman interpreted this disparity as the true cause of sexual selection; multiple mating afforded a much higher reproductive payoff to males than it did to females.

The slopes in sexual selection gradients (or Bateman gradients) are often used to quantify and compare the relative intensities of sexual selection operating on males and females of the same or different species (e.g., Arnold and Duvall 1994; Andersson and Iwasa 1996; Jones et al. 2000, 2002; Lorch et al. 2008). The Bateman gradient approach to quantifying sexual selection does have several qualifications and limitations (Parker and Tang-Martinez 2005; Tang-Martinez and Ryder, 2005; Snyder and Gowaty 2007). Nevertheless, for numerous separate-sex species, many researchers now accept the basic thrust of Bateman's argument: reproductive success in males tends to be limited by access to mates whereas reproductive success in females tends to be limited by access to resources (Jones and Ratterman 2009). None of this is to imply that females can never benefit from having multiple mates. Among the immediate benefits that a female might gain from multiple partners are more nuptial gifts, improved territorial access, or more help with parenting. A female might also benefit indirectly from multiple mating, via more or better quality offspring. If a female mates successfully with more than one male, genetic diversity in her clutch will be higher, she has more opportunities to obtain "good genes" for her progeny or to improve the genetic compatibility between

(continued)

BOX 1.5 *(continued)*

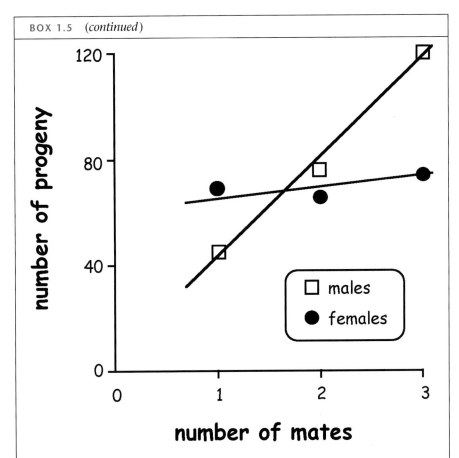

herself and some of her mates, and she in effect gains reproductive protection (fertilization insurance) against the possibility that some of her mates are sterile. However, none of these factors is likely to boost a female's genetic fitness nearly as much as multiple mating by a male normally can boost his genetic fitness. Of course, multiple mating by either sex can have major costs as well, such as the time and energy involved in finding and seducing mates and the increased likelihood of contracting sexually transmitted diseases.

Especially following Willson's (1979) landmark treatment of sexual selection in plants, Bateman's concepts have also been applied to many plant species. For example, if Bateman is strictly correct, then female fitness would not be pollen-limited and male fitness would be mate-limited. Empirical tests of these and various associated possibilities have been addressed widely in the botanical literature (e.g., Stanton et al. 1986; Campbell 1989; Morgan and Conner 2001; Knight et al. 2005).

cuckolded by a competitor male, natural selection often tends to disfavor the evolution of male acquiescence to mating with a single polyandrous partner.

(d) *greater postzygotic investments in offspring by females than by males.* In many species, anisogamy itself tends to predispose females more so than males to devote additional energy and resources to rearing progeny (because a female already has invested a greater down payment in the zygote). This disparity has become greatly exaggerated in many species, for example in all mammals and in various viviparous (live-bearing) fishes with internal female pregnancy. In such species, postzygotic care by females clearly is both mandatory and extensive. Postzygotic care of embryos is also characteristic of many separate-sex (dioecious) plant species in which females bear the burden of carrying the fertilized seeds. Female-biased parental care, combined with the backdrop of anisogamy, gives a synergistic boost to the disparity of reproductive life experiences between males and females in many separate-sex species.

(e) *higher potential variability in reproductive success among males than among females.* Especially in a highly polygynous species, some males can be huge winners in the reproductive sweepstakes while other males may fail completely. The result is a higher inter-male than inter-female variance in reproductive output (genetic fitness). Under monogamy, by contrast, inter-male and inter-female variations in reproductive success are identical.

(f) *greater competition for mates among males than among females.* This expectation likewise applies to polygynous species more so than to species that are primarily monogamous, and for reasons mentioned earlier it should also apply more strongly to males in polygynous systems than to females in polyandrous systems. This is not to imply, however, that males and females in strictly monogamous species are immune to sexual selection pressures related to mate choice and reproductive competition, because, under any mating system, each individual can improve its genetic fitness by choosing mate(s) of the highest possible quality.

(g) *greater opportunity for and strength of sexual selection on males than on females.* This follows directly from (f), and again applies especially to polygynous species.

(h) *greater elaboration of secondary sexual traits in males than in females.* This follows directly from (g) and also applies especially to polygynous species.

All of these extended ramifications of anisogamy are merely general evolutionary expectations, and none is a universal outcome in the highly heterogeneous biological world (Bonduriansky 2009). For one example, many invertebrate animals broadcast their eggs and sperm in the ocean, and empirical evidence suggests that sperm often can be limiting in such situations (Levitan and Petersen 1995). For another example, mate choice by males is documented to be commonplace in many animal species (Gwynne 1991), notwithstanding the general expected tendency for females to be the especially choosy gender. Furthermore, females sometimes can become adorned with secondary sexual traits too, especially in taxa with sex-role-reversal (in which sexual selection operates more intensely on females than on males). For example, in pipefish species (Syngnathidae), males receive eggs from females and, after fertilizing them, carry the embryos in an internal brood pouch during an extended male pregnancy. In many (but not all) pipefish species, the males' collective capacity to brood eggs is less than the females' total capacity to produce eggs (Vincent et al. 1992). This changes the basic evolutionary ground rules by making males—more so than females—the limiting factor in reproduction (notwithstanding the retention of anisogamy). Not surprisingly then, females in some pipefish species have higher variation in reproductive output than males, compete more intensely for mates, experience sexual selection more intensely than do males, and have become more adorned with secondary sexual characters (Jones and Avise 2001). These are all mirror-image outcomes compared to many other animals where females are the limiting resource in reproduction and thus tend to experience less intense sexual selection.

Nevertheless, each of the broad expectations listed above is met in a sufficiently wide variety of animal species as to make the occasional exceptions (such as in pipefishes, various marine invertebrates, and others) of special interest to evolutionary researchers. Finally, none of the standard ideas presented above preclude the likelihood that individuals of both sexes are sometimes under strong selection pressure with regard to mate choice; males as well as females are likely to improve their reproductive success by choosing high-quality partners. Indeed, theory as well as empirical data indicate that both males and females in many separate-sex taxa are likely to exhibit considerable adaptive flexibility in their mating decisions, depending on a wide range of ecological and behavioral factors (Gwynne and Simmons 1990; Gowaty and Hubbell 2009). For hermaphroditic species, such mating flexibility becomes even more catholic, because each individual in effect chooses whether to behave as a male or as a female (or perhaps both) at particular stages of life. Furthermore, dual-sex individuals in many hermaphroditic taxa can "decide" whether to self-fertilize or to outcross with another individual.

Sex Ratios in Separate-sex Taxa

The sex ratio in any separate-sex population is the relative number of males versus females. Depending on the context, sex ratio can refer to numbers at the time of conception (primary sex ratio), at the end of the period of parental care (secondary sex ratio), among independent non-breeders (tertiary sex ratio), or among reproductively active adults (the quaternary or breeding sex ratio). The concept of "operational sex ratio" (OSR) is also important. A population's OSR is the number of reproductive males versus females (or their gametes) effectively available during the time period under consideration. It can differ from the census sex ratio of a population for many reasons, such as the fact that individuals inevitably vary in relevant parameters such as age, fecundity, and reproductive availability. The OSR is often a better indicator (than the face-value number of males and females) of the nature and intensity of sexual selection likely to be at work in a given population.

Relationship to Sexual Selection and Mating Systems

Three centuries ago, an English physician to Her Majesty the Queen stated that a 1:1 sex ratio (an equal number of males and females in a population) provided a strong argument for Divine Providence: "for by this means it is provided that the Species shall never fail, since every Male shall have its Female, and of a Proportionable Age" (Arbuthnot 1710, quoted from Skyrms 1996). In a somewhat similar vein, Charles Darwin (1871) speculated that a 1:1 sex ratio is advantageous because it minimizes fighting over mates (although later in the same book he admitted that the sex-ratio problem "is so intricate that it is safer to leave its solution to the future"). Other biologists sometimes have assumed that a population would fare better with an excess of females because more total offspring then could be produced (if particular males service multiple females). A related notion was that extra males are reproductively superfluous and thus could be dispensed with. All of these interpretations naively invoke group selection: the generally invalid notion that natural selection operates for the collective good of a population or species, rather than by favoring particular individuals with higher genetic fitnesses.

In 1930, Ronald Fisher essentially solved the sex-ratio problem by elaborating a simple insight: in each and every generation in any sexual species, exactly 50% of all autosomal genes in progeny come from males (the sires) and 50% come from females (the dams). In other words, regardless of the sex ratio in a population, every individual has a father and a mother. With respect to sex ratio, this obvious truth has profound implications, which Fisher was the first to explicate in terms of how natural selection can operate at the level of alternative family tactics. He showed, mathematically, that a form of frequency-dependent selection acting on individual families ultimately would favor, at equilibrium, an equality of parental investment in sons and daughters. When males are in short supply in a population, particular families that produce son-biased litters would tend to get more genes into subsequent generations, and the frequency of males in the population would temporarily rise; conversely, when females are in short supply, families that produce daughter-biased litters would tend to have higher mean fitnesses, so females would tend to increase in frequency in the population, for a time (table 1.1).

Eventually an equilibrium is achieved wherein the realized sex ratio in a population comes to equal the relative parental costs of raising individual sons versus daughters. For example, if each daughter is twice as costly to produce as each son, the population sex ratio should stabilize at about 2 males : 1 female. If progeny of the two sexes are about equally expensive for parents to produce (as is true in many species), the population sex ratio

TABLE 1.1   Quantitative example of how natural selection acting on individual families tends to buffer a population's sex ratio against departures from a stable equilibrium point (after Pianka 1988). In this case, we assume that each son or daughter is equally expensive for parents to produce, and we consider the expected relative genetic contributions (C) to future generations of otherwise comparable families that invest differently in sons versus daughters.

		Number of Males	Number of Females
Case I.	Initial population	100	100
	family a	2	2
	family b	4	0
	Subsequent population (sum)	106	102
	$C_a = 2/106 + 2/102 = 0.03848$ (the slightly winning tactic)		
	$C_b = 4/106 + 0/102 = 0.03774$		
Case II.	Initial population	150	50
	family a	2	2
	family b	4	0
	Subsequent population (sum)	156	52
	$C_a = 2/156 + 2/52 = 0.05128$ (the strongly winning tactic)		
	$C_b = 4/156 + 0/52 = 0.02564$		
Case III.	Initial population	50	150
	family a	2	2
	family b	4	0
	Subsequent population (sum)	56	152
	$C_a = 2/56 + 2/152 = 0.04887$		
	$C_b = 4/56 + 0/152 = 0.07143$ (the strongly winning tactic)		

should stabilize at roughly 1:1, and the optimal family sex ratio should be 0.5. Any departures (but especially large ones) from this stable equilibrium point would then be countered by natural selection (fig. 1.4). In other words, a relative scarcity of either sex in a population would make increased production of that sex temporarily worthwhile for particular parents, and this would tend to redress the numerical imbalance between the two sexes.

Although Fisher's model does not fully apply under all conditions (such as when male reproductive variation is high; see chapter 3), it invokes no group selection and thus was a great intellectual advance. The model simply predicts the extended consequences of conventional selection operating on each family's genetic fitness, which in turn depends on its son:daughter ratio in comparison to the male:female ratio in the broader population. The equilibrium sex ratio under Fisher's model is an example of what John Maynard Smith (1976) has called an evolutionary stable strategy, or ESS. In general, an ESS is an evolutionary outcome (such as a population's sex ratio at equilibrium) that can resist permanent invasion by an alternate tactic (such as might be introduced by a new mutation or genetic recombination, for example). Mathematical approaches to identifying ESSs are standard practice in evolutionary game theory (box 1.6), where the usual goal is to illuminate the behavioral tactic(s) that benefit personal fitness, and to discover whether and under what circumstances a particular behavioral tactic or combination of tactics in a population is immune to invasion by alternative tactics. Each non-invasible outcome is an evolutionary stable strategy (ESS), or the situation toward which a population might tend to evolve (if the assumptions underlying the model are valid).

Fisher's sex-ratio theory provides an individual-selection explanation for population sex ratios, including the routine appearance in many species of population sex ratios near 1:1. One long-term outcome of such frequency-dependent selection on families presumably has been the evolutionary emergence of proximate sex-determining mechanisms, such as the familiar Mendelian systems of XY and ZW that produce, via the segregation rule of meiosis, approximately equal numbers of males and females in many separate-sex species. But in any system of genetic sex determination (GSD), and certainly in species with environmental sex determination (ESD), any pronounced population-level excesses of males or of females may temporarily renew selection pressures favoring suitable adjustments in how individual families invest in sons versus daughters (box 1.7). Whether such adjustments materialize will depend not only on the selection pressures but also on the nature of sex determination and whether suitable genetic variation exists for evolutionary responses (Uller et al. 2007).

As described in the previous section and summarized in figure 1.3, a population's "operational" sex ratio (how many reproductively competent males and females in effect are present; Emlen and Oring 1977) is an integral com-

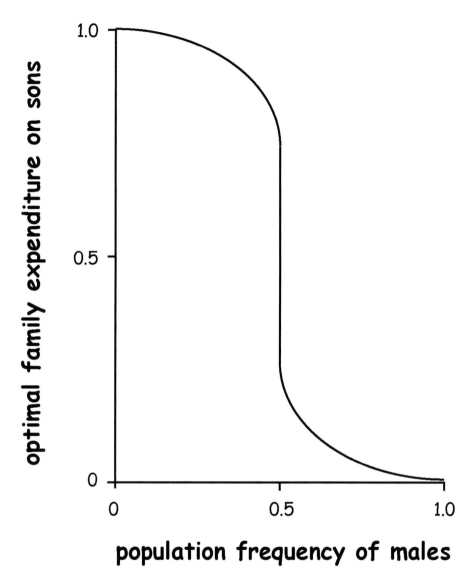

FIGURE 1.4   Frequency-dependent selection on individual families in a population in which offspring of either gender (sons or daughters) are equally expensive, per capita, for parents to produce (after Pianka 1988).

BOX 1.6   Evolutionary Game Theory

The behavioral tactic that optimizes an individual's genetic fitness often depends on what other individuals do. This realization led to the development of a sophisticated branch of population biology termed *game theory* (Maynard Smith 1982), which applies to many kinds of biological, economic, and social interactions (Skyrms 1996; Osborne 2004). Game theory analyzes how animals might act when an individual's success depends on others' decisions as well as its own. In human social behavior, applications include bargaining encounters and justice, ultimatum games, commitment choices, decisions about mutual aid versus defection, hawk-dove interactions, ownership issues, and truth versus deception in communication and signaling. Game theory also finds many applications in an animal's "decision-making" in nature. For example, should a family produce mostly sons, mostly daughters, or some mixed tactic (see text)? Or, can the strategy of hermaphroditism invade a separate-sex species? The usual goal in game theory is to illuminate the behavioral tactic(s) that benefit personal fitness, and to discover whether and under what circumstances particular tactics or combinations of tactics are evolutionarily stable (immune to invasion by alternative tactics) in a population. Each non-invasible outcome is an evolutionary stable strategy (ESS), or the situation toward which a population might tend to evolve (if the assumptions underlying the model are valid).

To assess whether at least one ESS exists, researchers usually begin by developing a payoff matrix showing the changes in fitness (i.e., the gains and losses in the contest) to an individual who adopts one or another specified tactic when its opponent adopts one of those tactics. For example, in a hawk-dove contest over limited resources, two doves might share the resource, a hawk takes all the resources from a dove, and two hawks fight and injure one another. Thus, the payoff matrix could be as follows (where the payoffs are to an individual adopting the tactic on the left if its opponent adopts the tactic listed above):

	hawk	*dove*
hawk	−2	2
dove	0	1

In general, the values in any payoff matrix rely on independent knowledge (or, sometimes, mere assumptions) about the situation. Knowing these payoffs and the initial frequencies of hawks and doves in a population, game theorists use a series of mathematical equations to quantify selection-mediated changes in the frequency of hawks and doves, genera-

(*continued*)

BOX 1.6 (*continued*)

tion-by-generation, and to determine whether a particular tactic or mixture of tactics is evolutionarily stable. In this particular case, it can be shown (see Maynard Smith 1998) that a population evolves to either of two ESS states: (a) 1/3 pure hawks and 2/3 pure doves; or (b) each individual playing hawk or dove with probabilities 1/3 and 2/3, respectively.

Game theory in its simplest form pertains to pairs of competing tactics in asexual populations (such that each offspring has the same tactic as its parent). For such relatively simple cases, two general conclusions are as follows: at least one ESS always exists; if more than one ESS exists, the population evolves to one or the other depending on its initial state (i.e., depending on the "basin of attraction"). However, if more than two pure tactics are possible, there sometimes is no ESS, and the population might evolve in a cyclical manner indefinitely. Additional complications often arise when considering continuously varying (rather than discrete) tactics, and sexual reproduction (especially when the genetic system cannot directly generate the specified distribution of tactics).

BOX 1.7 Adjustments in Sex Ratio

Any evolutionary response to selection pressures for altered sex ratios in a population is predicated on the availability of suitable genetic variation upon which natural selection can operate. For species with environmental sex determination (ESD)—such as TDSD (box 1.2)—an evolutionary response might entail changes in the frequencies of genotypes that underlie the magnitude or nature of phenotypic responsiveness to the relevant environmental cue(s). For example, TDSD could be converted partially or entirely to GSD if particular genotypes that otherwise are quiescent become mutated or reregulated in such a way as to assume a more active role in sex determination. Some fish species show evidence for standing genetic variation of this sort. Thus, in *Menidia* silversides, breeding studies have documented a genetic component to sex determination, notwithstanding a strong influence of temperature in particular populations (Conover and Kynard 1981); and, different populations of *Menidia menidia* show different net balances of GSD and ESD (Lagomarsino and Conover 1993).

For species with GSD, adaptive evolutionary shifts in family sex ratios would entail selection-mediated changes in the frequencies or expressions of sex-determining genotypes. This could happen by any of several mechanistic routes, such as basic changes in how sex is genetically determined,

(*continued*)

BOX 1.7   (*continued*)

or genetic modifications to existing sex-control systems. In species with an XY-system, for example, the ratio of X-chromosomes to Y-chromosomes in gametes could in principle be altered by meiotic-drive mutations that modify the segregation patterns during meiosis (Burt and Trivers 2006).

Facultative adjustments in sex ratio, in response to environmental conditions, are also possible, as has been demonstrated not only in particular plant species (e.g., López and Domínguez 2003) but also in some animal taxa such as birds with seemingly strict GSD (Emlen 1997). For example, female kakapo parrots (*Strigops habroptilus*) fed a rich diet produce significantly more sons than daughters (Clout et al. 2002). In Seychelles warblers (*Acrocephalus sechellensis*), females appear to adjust the sex of their eggs adaptively, in accord with territory quality, by producing mostly daughters (which later serve as helpers at the nest) when occupying good habitats and mostly sons (who emigrate) when occupying poor-quality habitats (Komdeur et al. 1997, 2002). And in Gouldian finches (*Erythrura gouldiae*), researchers discovered that "females paired with genetically incompatible males of alternative color morphs overproduce sons, presumably to reduce investment in inviable daughters" (Pryke and Griffith 2009:1605). All birds are thought to possess the ZW system of GSD, so exactly how females mechanistically might adjust the primary sex ratios in their progeny, facultatively, remains to be illuminated. (Facultative postzygotic shifts in family sex ratios are easier to rationalize because parents in principle could devote greater care to sons or to daughters if it genetically benefited them to do so in particular circumstances.)

ponent of a complex nexus of interrelated evolutionary factors that also includes anisogamy, the mating system, the intensity and direction of sexual selection with respect to gender, and the potential in males versus females for the elaboration of secondary sexual traits (including behavioral as well as morphological characters). These and related factors are all important when considering the reproductive selective pressures at work in any separate-sex species.

Impact on Demographic Parameters

Apart from its key position in theories of mating systems and sexual selection, sex ratio also can influence important population genetic parameters (Hartl and Clark 1997). For example, a separate-sex population with a consistently male-biased or female-biased sex ratio has a much smaller effective population size than an otherwise comparable population composed of equal numbers of males and females. The effective size of a population (N_e) refers to the number of individuals in an idealized population displaying the same genetic properties (such as the inter-generational variance in allele frequencies due to genetic drift) as those observed in an actual population under consideration (Wright 1931; Charlesworth 2009). Usually, N_e is much smaller than N (the census size of a population) for one or more of the reasons discussed in box 1.8.

BOX 1.8   Why Effective Population Size Is Often Much Smaller than Census Population Size

(a) *Fluctuations in population size.* Populations in nature routinely fluctuate in size due to diseases, changes in habitat quality, shifting climates, predation, etc. The effective population size due to such fluctuations is equal to the harmonic mean of breeding population sizes across generations. A harmonic mean is a function of the mean of reciprocals, or in this case

$$N_e = n / [\Sigma\ (1/N_i)], \qquad (1)$$

where N_i is the population size in the ith generation and n is the number of generations. A harmonic mean is closer to the smaller than it is to the larger of a series of numbers, so N_e can be much lower than most population censuses. A severe reduction in population size is called a population bottleneck, which can greatly depress a population's evolutionary effective size.

(b) *Variation in progeny numbers.* Even in a non-fluctuating population, some individuals may leave many more progeny than others, creating a large fitness variance across families. Only when offspring numbers follow a Poisson distribution with mean (and hence variance) of 2.0 per family, does $N_e = N$. In more realistic situations, where the variance often exceeds the mean, N_e is smaller than the census-breeding population size (Crow 1954). Organisms with extremely high fecundity are especially prone to disparities between N_e and N due to a high variability in fecundity across individuals (e.g., Hedgecock et al. 1992).

(c) *Extinction in subdivided populations.* In a species composed of many subpopulations each of which is subject to periodic extinction and

(continued)

BOX 1.8   (*continued*)

---

recolonization, the species as a whole can have a much lower N_e than otherwise indicated had only census sizes for specific generations been available (Maruyama and Kimura 1980).

(d) *Unequal sex ratio.* Sex ratio can also have a significant impact on N_e. In many separate-sex species, one gender may be more common than the other, if only periodically or temporarily. Let N_m and N_f be census numbers of adult males and females. The effective population size due to any disparity between N_m and N_f can be calculated as

$$N_e = 4N_m N_f / (N_m + N_f). \tag{2}$$

This equation shows that N_e is less than the total census count ($N_m + N_f$) in all cases except when $N_m = N_f$. In effect, the rarer sex provides a partial bottleneck (relative to the common sex) through which gametes must squeeze in passing from one generation to the next.

(e) *Joint effects of multiple factors.* If the census population sizes of males and females are known across multiple generations, then the joint effects of sex ratio and population fluctuations on N_e can be determined by converting the census sizes of males and females to an effective population size for that generation (using equation 2), and then calculating the harmonic mean of these single-generation estimates (using equation 1).

Sex Ratios in Hermaphroditic Taxa

Most of our discussion to this point has focused on gonochoristic or dioecious (separate-sex) species. A recurring set of questions throughout this book is whether (and if so, how) the theories outlined in the preceding sections might apply to hermaphroditic species. One issue related to anisogamy can be clarified at the outset. By definition, any hermaphroditic individual reproduces as a male when it contributes the smaller-size gamete to a zygote; and it reproduces as a female when it contributes a larger-size gamete. In this important sense, anisogamy applies with full force to hermaphroditic species, except that the bimodal distribution of gamete size occurs within each hermaphroditic individual rather than between separate males and females. Similarly, when referring to relative investments of a hermaphrodite in male versus female functions, we can speak, by definition, of how the individual allocates its finite energy and resources to producing small versus large gametes.

As applied to hermaphrodites, sex allocation theory (discussed further below) describes how dual-sex organisms should optimally divide their resources between male and female reproductive functions at various stages of life. These "decisions" by individuals will also have several ramifications for the population's overall sex ratio (box 1.9). This is so because when an individual "decides" to invest exclusively in small-size gametes, in effect it has chosen to be a male, and when it "decides" to produce large-size gametes exclusively, it becomes in effect a female. In many hermaphroditic species,

---

BOX 1.9   Sex-ratio Evolution in Sex-changing Organisms

As detailed by Allsop and West (2004), sex allocation theory makes at least three clear predictions about the overall sex ratio in hermaphroditic species in which individuals have the capacity to change sex. First, such organisms should have an overall population sex ratio that is biased toward the "first" or initial sex. Second, this bias should be less extreme in partially sex-changing organisms, where a proportion of the "second" sex matures directly from the juvenile stage. Third, the sex ratio should be more biased in protogynous (female-first) species than in protandrous (male-first) species. Allsop and West (2004) tested these predictions with a comparative literature review on 121 species of sex-changing animals spanning five phyla (fish, arthropods, echinoderms, mollusks, and annelids). They found strong support for the first and third predictions, but only partial support for the second.

To illustrate the kinds of logic underlying these predictions, consider the argument for prediction 1 as presented by Allsop and West (2004; following Charnov 1993): "Consider the case of a protogynous diploid species, in which individuals mature as females and then change sex to males when older (bigger). In this case the relative fitness of males increases faster with age than it does for females. Males and females must make an equal genetic contribution to the next generation, because all offspring have two parents. Consequently, it must be true that $N_m W_m = N_f W_f$ (equation 1), where N_m and N_f are the number of mature males and females and W_m and W_f represent the reproductive value (fitness) of a male and a female. Given that the reproductive value of a male at the point of sex change will be equal to that of a female, and that reproductive value increases faster with age, it follows that $W_m > W_f$, because any individual that has become a male must have a higher fitness than individuals that are still female. Consequently, for equation 1 to hold, it also follows that $N_m < N_f$. This means that there will be more females than males, and hence a female-biased sex ratio. The converse prediction for protandrous (male-first) species can equally be made, showing that a male-biased sex ratio is predicted."

each individual may have the capacity to switch its functional gender during its lifetime, and the collective actions of many such individuals can affect the population's sex ratio.

Because anisogamy and sex-ratio concepts apply to hermaphroditic as well as separate-sex taxa, it should probably come as no great surprise that the theories previously developed for many other interrelated reproductive topics (including mating systems, sexual selection, and sexual dimorphism) can similarly apply, with appropriate modifications, to hermaphroditic as well as to dioecious taxa.

Hermaphroditism Versus Separate Sexes

Early researchers sometimes invoked population-level advantages to rationalize the evolution of hermaphroditism. For example, Moe (1969) suggested that sequential hermaphroditism might have evolved as a population-control mechanism, with the age of transformation between female and male shifting up or down to compensate for whether a population was too sparse or too dense. Other hypotheses with a group-selection aura posited that hermaphroditism is favored routinely during the evolutionary process because it might increase total zygotic production in a population (Smith 1967) or perhaps focus sexual performances into age classes that would maximize a population's reproductive output (Nikolski 1963). By contrast, more modern views have emphasized how natural selection and sexual selection might operate on the *differential fitnesses of individuals*, in various social or ecological contexts, to influence the evolution of various forms of hermaphroditism.

Fitness Considerations

Hermaphroditism and separate-sexes can be interpreted as competing tactics in evolutionary game theory. An important question then becomes: Under what conditions might a gene for hermaphroditism invade a separate-sex population (or perhaps *vice versa* in some cases)? The theoretical answer is straightforward in broad outline (Maynard Smith 1998). In a separate-sex population with an equilibrium 1:1 sex ratio and in which males and females have equal reproductive success (R_m and R_f, respectively), assume that any *de novo* hermaphrodites outcross only. Let αR_m and βR_f be the reproductive successes of hermaphrodites as males and as females, respectively. Hermaphroditism can then invade the population if $\alpha + \beta > 1$. In other words, the separate-sex condition is susceptible to invasion by hermaphrodites when the average reproductive success of the latter exceeds what would be expected for standard individuals that reproduce solely as a male or a fe-

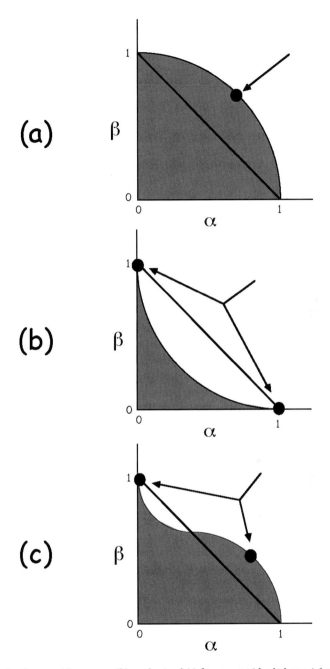

FIGURE 1.5 Convex (a), concave (b), and mixed (c) fitness sets (shaded areas) for hermaphrodites (after Maynard Smith 1998). α and β represent the reproductive successes of hermaphrodites functioning as males and as females, respectively, relative to their separate-sex counterparts. These sets describe the general fitness conditions in which hermaphroditism (a), separate-sexes (b), or mixtures of the two (c) are predicted to be evolutionarily stable (see text). Arrows and black dots indicate these various outcomes.

male. However, α and β may vary if different hermaphrodites invest differentially in male versus female reproductive functions. Figure 1.5 shows fitness sets for all possible pairs of α and β values in hermaphrodites. If the set is convex (fig. 1.5a), hermaphroditism is the ESS; if the set is concave (fig. 1.5b), the ESS is separate-sexes. If the fitness set is a mixture of convex and concave (fig. 1.5c), then a mixed population of single-sex individuals and hermaphrodites may be an ESS (Charnov et al. 1976).

For biologists, the interesting questions then become what ecological, phenotypic, or other considerations predict a convex versus a concave fitness set for hermaphrodites. Many factors can come into play. For example, in plants a hermaphrodite's expenditure on showy petals and nectar can serve both male and female functions (attracting pollinators), thereby economizing on resources in ways that might tend to promote a convex fitness set, all else being equal. Indeed, hermaphroditism seems to be more common in animal-pollinated plant species (where showy flowers are often needed to attract pollinators) than in wind-pollinated plants. (However, many other factors might also explain this relationship, such as the idea that separate sexes are more often needed to prevent inbreeding in wind-pollinated species; or that for wind-pollinated species, a plant's investment in male function is not likely to lead to diminishing returns the way it may with insect-pollinated species.) Conversely, many vertebrate animals have sexual organs and sexually selected traits that probably would be ineffective if they were developed to only a partial extent in hermaphrodites, thus promoting a concave fitness set, all else being equal. Such considerations might help to account for the observation that hermaphroditism is common in plants but relatively rare in vertebrate animals. Another factor likely to be associated with a convex fitness set is a low density of conspecific individuals, in which case hermaphroditism may be selected because of the fertilization advantages it can confer (Charnov et al. 1976). Specifically, an outcrossing hermaphroditic individual need encounter only one other individual (rather than an individual of the proper sex) in order to mate, and a self-fertilizing hermaphrodite need encounter no one else to reproduce successfully via selfing.

Mating Systems and Related Phenomena

Hermaphrodites raise many additional reproductive issues that extend and sometimes challenge the kinds of theories outlined above for separate-sex taxa. What kinds of mating systems do hermaphroditic species display, and how might these relate to ecological or evolutionary circumstances? Does sexual selection operate in hermaphroditic species, and if so, how does it vary in intensity and direction vis-à-vis alternative mating systems? Can hermaphroditic species display pronounced dimorphism in secondary sexual traits (as do many separate-sex species), and if so how, when, and why?

How does the concept of sex ratio (and its extended population-genetic ramifications, such as effective population size) apply to dual-sex species? How successful in both the short-term and long-term are hermaphroditic species compared to their separate-sex counterparts? These and other ecological and evolutionary questions about hermaphroditism long have fascinated biologists.

One mating-system topic—selfing versus outcrossing—is uniquely relevant to hermaphroditic taxa. Selfing (self-fertilization) entails, in effect, mating with oneself; a single individual produces both the male and female gametes that unite to form a zygote. Clearly, selfing is not possible in separate-sex taxa (although other routes to inbreeding exist via matings among relatives), but hermaphrodites in many dual-sex species engage routinely in self-fertilization (an especially intense form of inbreeding). In populations of many hermaphroditic species, individuals may either outcross (mate with separate individuals) exclusively, or they may display a mixed-mating system (Clegg 1980; Brown 1989) with intermediate frequencies of selfing and outcrossing.

Self-fertilization is an extreme form of incest (sex with close genetic kin). Especially when selfing extends across successive generations, genetic variation (heterozygosity) within each genetic lineage quickly falls (fig. 1.6) and each inbred line soon becomes highly homozygous. This also means that genetic recombination in effect is suppressed in inbred lineages (despite the retention of meiosis and syngamy) because little genetic variation exists to be shuffled during sexual reproduction into novel multilocus combinations (Allard 1975).

Many plant and animal species display inbreeding depression (Ralls and Ballou 1983; Crnokrak and Roff 1999; Frankham et al. 2002): diminished survival or fertility associated with matings between genetic relatives. The reductions in genetic fitness are sometimes high (50% or more) but also highly variable among species. Two hypotheses for inbreeding depression have been debated extensively (Charlesworth and Charlesworth 1987b). Under the *dominance* hypothesis, lowered fitness results from particular loci becoming homozygous for rare deleterious recessive alleles whose expression in outbred populations is masked in heterozygotes. Under the *overdominance* hypothesis, a genome-wide drop in heterozygosity is the causal factor for the fitness decline. The dominance hypothesis also implies that if a population survives an initial bout of intense inbreeding, it might then persist indefinitely (or at least until the environment changes beyond the population's genetic scope) because natural selection has purged the genome of deleterious recessive alleles. By contrast, the overdominance hypothesis predicts that an inbred population, if it survives, will continue to perform poorly because its heterozygosity remains low.

Recent literature offers considerable support for the dominance model (at least in plants), with overdominance playing a secondary but still important

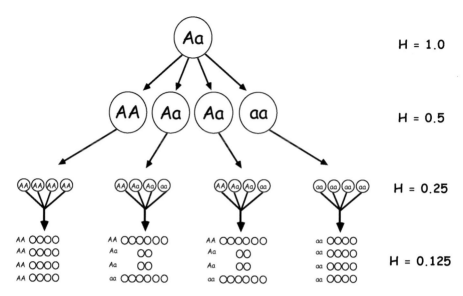

FIGURE 1.6   The rapid loss of heterozygosity (*H*) under selfing, in this case across four successive generations (from Avise 2009). For any autosomal gene that is heterozygous (Aa) initially, a self-fertilizing hermaphrodite produces offspring in an expected Mendelian ratio of 1(AA):2(Aa):1(aa), that is, 50% homozygotes and 50% heterozygotes. Each offspring that is homozygous and self-fertilizes produces only homozygous progeny, and each self-fertilizing offspring that is heterozygous again produces homozygous and heterozygous progeny in a 1:1 ratio. The net result is a precipitous decline in heterozygosity (by 50% per generation) in any population of self-fertilizing hermaphrodites.

role (e.g., Carr and Dudash 2003). Regardless of the mechanism underlying inbreeding depression, the phenomenon motivates many questions about the evolutionary histories and the adaptive significance of selfing versus outcrossing in various hermaphroditic taxa. Indeed, much of the genetic literature on simultaneous hermaphrodites tends to regard selfing versus outcrossing as a primary consideration when discussing genetic fitnesses under alternative mating systems. The separate-sex condition itself can be viewed as an outcross-enforcing mechanism that sometimes may have invaded an ancestral population of hermaphrodites via selection against selfing.

Sex Allocation

In 1982, Eric Charnov published an important book that helped to extend the "selection thinking" promoted by Ronald Fisher (1930), Maynard Smith (1978), and others (Williams 1966) to the topic of sex allocation (SA) in hermaphrodites (see also Charlesworth and Charlesworth 1981). Charnov defined SA as the relative parental investment in male versus female reproductive functions, and he recognized a close analogy between the question of how a hermaphroditic individual allocates resources to male versus female gametes and how separate-sex parents allocate resources to produce sons versus daughters (Hardy 2002). The basic Fisherian truth holds in both cases: that each individual (unisexual or hermaphroditic) comes from a union of male gamete and female gamete. Thus, by analogy to the sex-ratio theory for separate-sex species, a hermaphrodite should be able to enhance its personal fitness by investing disproportionately in whatever type of sex cell (male or female) is otherwise in limited supply for fertilizations in the relevant pool of gametes. (For obligate outcrossers, the relevant gamete pool might be that of the local population or deme, whereas for selfers it could be the available pool of gametes within each individual.) Indeed, in making such SA decisions, hermaphrodites in general should show exceptional flexibility (relative to most gonochorists) because they need only allot different amounts of resources to different sexual functions or behaviors that (almost by definition) are already within their developmental and physiological repertoires (Hamilton 1967; Michiels 1998). Another distinction is that decisions about SA affect the genetic fitness of hermaphroditic individuals directly, whereas they affect the genetic fitness of unisex individuals via the reproductive successes of their sons and daughters (Borgia and Blick 1981; Michiels et al. 1999).

Most of the available SA theory for hermaphrodites can be interpreted in either of two contexts: as evolutionary models in which SA patterns evolve across generations, and/or as models of phenotypic plasticity wherein individuals make intra-lifetime adjustments in SA in response to short-term changes in population conditions (Schärer 2009; Schärer and Janicke 2009). Formal SA models usually include several simplifying assumptions, such as that all individuals in the population have a fixed resource budget for reproduction, and that there is a linear tradeoff between the allocations to male and female functions. Even so, theory and data indicate that many outcomes are possible, depending on the particular biological settings or conditions.

For example, some hermaphrodites produce male and female gametes simultaneously whereas others do so at different stages of life. Among simultaneous hermaphrodites, some species self-fertilize primarily, others outcross exclusively, and some have a mixed-mating system of selfing plus outcrossing. Thus, several ecological and behavioral factors can come into play that complicate predictions about what natural selection and sexual selection should perceive to be an individual's optimal reproductive tactic. Genetic considerations can be important too, including the topic of inbreeding depression in hermaphroditic taxa that self-fertilize.

Among sequential hermaphrodites, outcrossing necessarily is forced, but questions remain about when best in life to function as a male, when to function as a female, and when in life to make the optimal switch. The answers depend on many factors related to species-specific ecologies, genetics, behaviors, and life-history parameters. In searching for a common-denominator consideration, Ghiselin (1969:44) offered an important insight: "Suppose that the reproductive functions of one sex were better discharged by a small animal, or those of the other sex by a large one. An animal which, as it grew, assumed the sex advantageous to its current size would thereby increase its reproductive potential." Most organisms grow in size during their lifetimes (some more or less indefinitely, as in fish with indeterminate growth). Thus, the optimal reproductive tactic for a young individual may be very different from that of an old one, and organisms that are sequential hermaphrodites should be able to capitalize upon this fact, at least in principle. Ghiselin's size-advantage hypothesis remains a popular theme for addressing various expressions of sequential hermaphroditism in fishes and other animals, and the concept also has been applied to plants (Policansky 1982; Klinkhamer et al. 1997; Klinkhamer and de Jong 2002).

Phylogenetic Legacy Versus Contemporary Adaptive Significance

Most of the evolutionary theories and models introduced in the preceding sections (such as the ESS and SA approaches, the size-advantage hypothesis, the attempts to account for evolutionary conversions between separate-sexuality and hermaphroditism, or the attempts to analyze the benefits of outcrossing versus selfing) reflect adaptationist perspectives on sexual systems. In other words, they assume that the biological outcome in each extant lineage registers primarily the operation of natural selection or sexual selection. However, the present-day distribution of alternative sexual systems in various organisms undoubtedly represents some combination of contemporary adaptive significance and phylogenetic legacy. Teasing these influences apart and addressing their relative impacts are central challenges for biologists.

A functional approach is to assess genetic fitness as it relates to reproductive mode, for example by examining fecundity or ecological success in suitable experimental settings or from field observations. This can sometimes yield insights about what reproductive systems work best in particular species or ecological arenas. Another general approach involves the "comparative method" that seeks to draw conclusions from the current-day distributions of reproductive patterns across many species or higher taxa. This method typically entails compiling information on sexual systems (for example) from the scientific literature, and then interpreting, in a phylogenetic framework, the distributions of such traits across taxa (Harvey et al. 1996; Avise 2006).

The word *phylogeny* (from Greek roots "phyl" meaning tribe or kind, and "geny" meaning origin) refers to the evolutionary history of genealogical connections between ancestors and descendents. For the first century following Darwin, biologists estimated phylogenies for various lineages by comparing organismal phenotypes: morphological, physiological, or behavioral characteristics. For the last 50 years, many scientists have used molecular data (e.g., DNA sequences) to assess more directly the "genetic" component of phylogenetics. However, a molecular phylogeny usually remains of special interest when interpreted jointly with phenotypic data. In particular, a popular approach in the last 20 years has been to interpret the evolution of different organismal phenotypes (such as alternative sexual systems) against the historical backdrop provided by molecular phylogenies. The approach is called phylogenetic character mapping or PCM (box 1.10), and it can be applied to the alternative states of any genetically based phenotypic character.

Adaptive and phylogenetic explanations for the current distributions of phenotypes are seldom mutually exclusive (because natural and sexual selection certainly can impact the phylogenetic distributions of traits). Nevertheless, PCM can help to identify cases where a solid statistical case can (or cannot) be made for adaptive scenarios from comparative data. For example, suppose that for 12 surveyed animal species a perfect correlation is observed between sexual system (separate-sexes versus hermaphroditism in this case) and habitat type (marine versus freshwater). The interpretation of this association can differ depending on the species phylogeny. If these species are related as shown in the top half of figure 1.7, then the correlation is statistically and evolutionarily significant because, in all six independent phylogenetic contrasts, hermaphroditism was associated consistently with the marine habitat. Thus, it might be appropriate to propose that something about the marine setting predisposes for selection pressures favoring hermaphroditism (or, conversely, that something about the freshwater environment consistently selects for the separate-sex condition). On the other hand, if these species are related as shown in the bottom half of the figure, then the correlation is statistically nonsignificant (after correcting for phylogeny) because there might have been only one evolutionary event (e.g., at point X) that eventuated in the perfect association of hermaphroditism with the marine environment. If so, it would be premature to conclude from this evidence alone that hermaphroditism is more likely to evolve in marine than in freshwater settings.

Phylogenetic considerations also forcefully remind us that alternative adaptive tactics cannot be viewed as evolutionarily non-interconvertible. Switches between different sexual systems do occur, and they must involve intermediate stages. For hermaphroditism and separate-sexuality (gonochorism or dioecism), potential intermediate stages are gynodioecy (mixtures of

BOX 1.10   Phylogenetic Character Mapping (PCM)

PCM entails estimating the evolutionary histories of phenotypic characters or traits, such as different reproductive anatomies or sexual systems. A popular approach in recent years is to use an independent molecular phylogeny as backdrop to deduce ancestral and derived character states and to map or "reconstruct" evolutionary transformations among those character states on a tree. The process involves four basic steps: (a) gather extensive molecular data (using DNA-sequencing or other laboratory methods) from homologous genes in living species in a taxonomic group of interest; (b) apply appropriate phylogenetic algorithms to those genetic data to estimate a molecular phylogeny for those species; (c) survey and score the extant species for the variable phenotypes of interest (such as hermaphroditism versus separate-sexes); and (d) after plotting those character states on terminal (extant) nodes of the tree, use suitable phylogenetic methods to deduce the probable character states at various internal nodes and branches in the tree. From such PCM exercises, robust conclusions often can be reached about the evolutionary histories of phenotypes.

Although PCM is simple in principle, in practice there can be many complications and limitations related to the following types of questions: How accurate is the molecular tree? Are the phenotypes adequately described? And were the phylogenetic reconstructions properly conducted? Thus, PCM-based conclusions (like all conclusions in science) should be interpreted as reasoned but provisional hypotheses, always subject to reinterpretation with new or improved evidence.

The basic concept of PCM is outlined in the accompanying figure (from Avise 2006). Shown across the top are eight hypothetical species (A–H) displaying one or the other of two character states (white or black squares) of a phenotype. Knowledge of the evolutionary relationships of these species (e.g., from molecular genetic data) can be used to assess how these character states likely evolved. For example, if species A–H prove to be phylogenetically related as shown in diagram I, then white-square was probably the original ancestral condition for the group and black-square is a shared-derived condition (i.e., a synapomorphy) for the ADE clade. However, if the species are allied as shown in diagram II, then black-square was probably the ancestral condition and white-square is a shared-derived state for the clade BCFGH. Many other outcomes are possible. For example, if the true phylogeny for species A–H is as diagrammed in III, then white-square was probably the ancestral state from which black-square evolved independently on three separate occasions.

(continued)

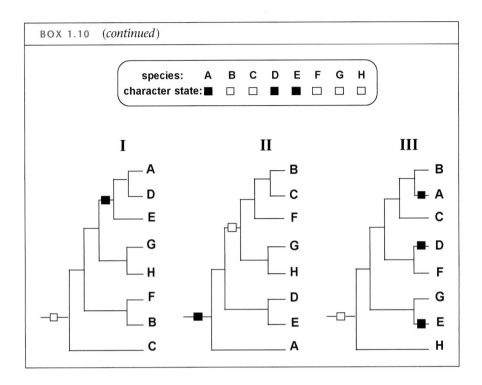

hermaphrodites and pure females within a species), androdioecy (mixtures of hermaphrodites and males), and trioecy (mixtures of hermaphrodites, males, and females), any of which themselves might be evolutionarily stable in some circumstances. Finally, when trying to interpret alternative reproductive strategies, it should always be remembered that evolutionary pathways in nature seldom are free to explore the entire terrain of a fitness landscape (Avise 2006). Instead, they follow trajectories that are constrained not just by contemporary selection but also by historical circumstance and by the available genetic variation upon which natural selection can operate. Such phylogenetic constraints mean that species do not always (and perhaps seldom) evolve theoretically optimal or even near-optimal reproductive solutions to nature's challenges.

With the introductory background provided in this chapter on the meaning of the two genders, and on alternative mating systems, sex roles, and related sexual topics, we are now ready to examine hermaphroditic plants, invertebrate animals, and fishes in greater detail. As we shall see in the ensuing chapters, the phenomenon of dual sexuality within the individual comes in many fascinating guises.

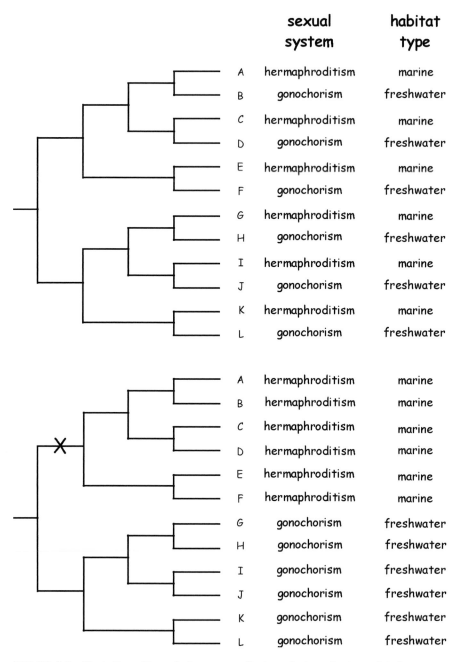

	sexual system	habitat type
A	hermaphroditism	marine
B	gonochorism	freshwater
C	hermaphroditism	marine
D	gonochorism	freshwater
E	hermaphroditism	marine
F	gonochorism	freshwater
G	hermaphroditism	marine
H	gonochorism	freshwater
I	hermaphroditism	marine
J	gonochorism	freshwater
K	hermaphroditism	marine
L	gonochorism	freshwater
A	hermaphroditism	marine
B	hermaphroditism	marine
C	hermaphroditism	marine
D	hermaphroditism	marine
E	hermaphroditism	marine
F	hermaphroditism	marine
G	gonochorism	freshwater
H	gonochorism	freshwater
I	gonochorism	freshwater
J	gonochorism	freshwater
K	gonochorism	freshwater
L	gonochorism	freshwater

FIGURE 1.7 Illustration of how phylogeny can affect conclusions about correlated biological variables (after Avise 2006). In both of the phylogenetic diagrams, sexual system and habitat type are perfectly correlated among the 12 extant species (A–L), but the statistical and biological interpretations of these data differ depending on how these species are phylogenetically related (see text).

SUMMARY

1. A hermaphrodite is a dual-sex organism that has the capacity to reproduce as both male and female during its lifetime. Hermaphroditism is relatively rare in vertebrates (being confined to several hundred species of fish), but extremely common in plants and invertebrate animals. Hermaphroditic species are sexual (as opposed to clonal or asexual), but they are to be distinguished from separate-sex species (gonochoristic animals or dioecious plants) that also reproduce sexually but consist of distinct male and female specimens.

2. Anisogamy refers to the bimodal distribution of gametic sizes that characterizes all multicellular sexual species; by definition, male gametes are relatively small and female gametes are relatively large. Anisogamy arose early in the history of multicellular life, probably via a combination of evolutionary factors including disruptive selection on gametic size, selection against sexual diseases housed in the cell cytoplasm, and the resolution of potential evolutionary conflicts between nuclear and cytoplasmic genomes. Anisogamy likewise applies to hermaphrodites, except that in this case an individual can produce both male and female gametic types.

3. Anisogamy implies that males have an inherent capacity to produce large numbers of small and energetically cheap gametes whereas females can produce far fewer and individually more expensive eggs. In many separate-sex lineages, this basic sexual asymmetry has initiated an evolutionary cascade of effects that quite often includes a higher fecundity potential in males than in females, behavioral tendencies in males (more so than females) to seek multiple mates, a greater inclination toward polygyny than toward polyandry, greater investment by females in postzygotic care of progeny, higher variation in reproductive success among males than among females, greater competition for mates among males than among females, stronger opportunities for sexual selection on males than on females, and the elaboration of secondary sexual traits especially in males. Exceptions to these trends often provide special insights into the general relationships among mating systems, sexual selection, and sexual dimorphism. Important questions for this book are how these interrelated phenomena play out in hermaphroditic species.

4. Sex ratio (the number of males versus the number of females in a population) is also tied closely to the topics of mating systems and sexual selection. In the 1930s, Ronald Fisher developed a powerful theory for sex-ratio evolution that considers the fitness consequences to individuals of producing different son:daughter ratios (this theory quickly superseded group-selection arguments that formerly had been popular). In quite analogous fashion, a sex allocation theory was developed in the 1980s that can be applied to the

related question of how hermaphroditic individuals might optimally invest in producing male versus female gametes. In traditional sex-ratio theory and in sex allocation theory alike, the optimal reproductive tactics of individuals can depend on many ecological considerations and population-level parameters.

5. Reproductive tactics displayed by species in nature may not always reflect optimal solutions to ecological circumstance, nor are they necessarily evolutionarily stable. Instead, the current distribution of alternative sexual systems in various organisms probably represents some combination of contemporary adaptive significance and phylogenetic legacy (historical inertia and evolutionary constraints). Scientific exercises in "phylogenetic character mapping," in which alternative traits are plotted and ancestral states are reconstructed in a phylogenetic framework, can help to tease apart the relative contributions of contemporary and historical processes on the distributions of particular biological features in extant organisms.

CHAPTER TWO

Dual-sex Plants

Hermaphroditism in plants is a vast subject. This chapter and the next will highlight dual-sex phenomena in plants and invertebrate animals, respectively, and thereby introduce reproductive topics that these organisms have motivated in the scientific literature. My goals are to profile dual-sex plants and invertebrates in nature as fascinating creatures in their own right, provide a biological overview of various reproductive modalities in these organisms, and thereby also add empirical and conceptual backdrop for discussions of hermaphroditic vertebrates in chapter 4.

Botanists have long known that most species of angiosperms (flowering plants with fruit-encased seeds) include dual-sex individuals. For example, Darwin (1876, 1877) wrote extensively about the sexual anatomies and breeding modes of hermaphroditic plants; and an early 20th-century survey of floral anatomies in nearly 122,000 angiosperms revealed that more than 95% of these species consist in whole or in part of dual-sex specimens (Yampolsky and Yampolsky 1922). Indeed, in the latter survey, only 4% of the species were strictly dioecious (i.e., separate sexes), whereas 72% were composed entirely of dual-sex individuals with both male and female anatomies (and, presumably, dual-sex functions). Although many additional examples of dioecy later were uncovered in angiosperms (Bawa 1980; Givnish 1982; Maurice et al. 1993; Renner and Ricklefs 1995), a strict separate-sex condition remains far less common than dual sexuality. The latter is also common in gymnosperms (plants with naked seeds, not enclosed in a fruit), especially in trees with wind-dispersed seeds (Givnish 1980). For seed plants generally, the standard wisdom is that one form or another of dual sexuality was an ancestral state from which the condition of dioecy (separate males and females) evolved on multiple independent occasions (Charlesworth and Charlesworth 1978; Charlesworth 1985; Delph and Wolf 2005). The meristematic nature of plants, wherein the growth of reproductive tissues ensues from undifferentiated but mitotically active soma (rather than from a strictly sequestered germline), undoubtedly contributes to the extreme diversity and ontogenetic flexibility of sexual systems in plants.

Terminology

Before proceeding, some terminology must be clarified. In the botanical literature, plethoras of terms exist (and sometimes have been applied inconsistently) to describe various anatomical or functional sexual systems in plants (Bawa and Beach 1981; Charnov 1982; Sakai and Weller 1999). Dioecy refers to separate-sex species in which male (staminate) and female (pistillate) flowers (fig. 2.1) are housed in separate individuals. Thus, dioecy in plants equates to gonochorism in animals, and I will use the two words interchangeably. Botanists traditionally reserve the term *hermaphroditism* (in the narrow or strict sense) for plant species in which each individual shows bisexual or "perfect" flowers that by definition have both male and female structures (fig. 2.1); whereas the term *monoecy* is applied to plant species in which distinct male and female flowers consistently co-occur on each dual-sex specimen. The broader term *cosexual* is then sometimes applied to individual plants that have both male and female functions regardless of their particular floral arrangements (Lloyd 1984). Thus, both hermaphroditism and monoecy clearly qualify as reproductive modalities involving dual-sex individuals. In addition, some plant species have polymorphic combinations of dual-sex and unisex individuals. In a sexual system known as androdioecy, any one specimen displays either male or bisexual flowers; and in gynodioecy, any one specimen displays either female or bisexual flowers. In still other plant species, various combinations of dual-sex and unisex flowers reside on an individual plant. In the sexual system known as andromonoecy, most individuals bear both male and bisexual flowers; and in gynomonoecy, most individuals bear both female and bisexual flowers. Thus, altogether, at least six major categories of sexual systems in plants include at least some dual-sex individuals. These systems are diagrammed in figure 2.2. To complicate matters further, however, the relative proportions of male flowers, female flowers, and bisexual flowers actually can vary in different plant species along continua extending from near-zero to near-100%, both among populations and within individuals. For example, even in dual-sex plants with perfect flowers, some individuals may function mostly as male and others more so as female (Campbell 1998).

The botanical definitions in the preceding paragraph traditionally were based mostly on flower anatomy, but sexual systems in plants can also be

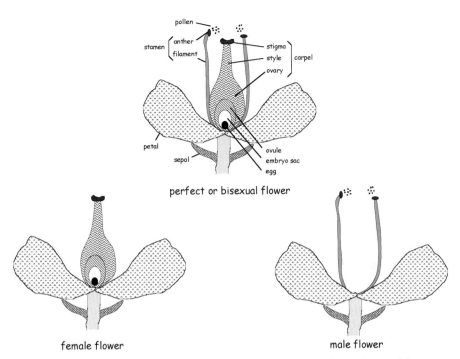

FIGURE 2.1   General flower types in angiosperm plants. Each perfect or bisexual flower has both male parts (the stamens, containing the pollen grains) *and* female parts (the pistil, made of one or more carpels including the egg, embryo sac, and the ovules). Each imperfect or unisexual flower has only functional male or female parts, although other vestigial components may be present. Especially in species that are animal-pollinated, male and female flowers as well as those that are perfect are all likely to have showy petals to attract pollinators, because the full pollination process entails both the donation and the receipt of pollen. The definitions and detailed knowledge of alternative flower types in conspecific plants trace back at least to Darwin (1877).

defined with respect to their realized reproductive functions (Lloyd (1972, 1979a; Primack and Lloyd 1980). Anatomical and functional considerations do not always agree. For example, some flowers that appear structurally hermaphroditic may produce only functional pollen or seeds (Lloyd and Myall 1976). Thus, even in species that are hermaphroditic in the narrow sense, some individuals may act more like males and others may perform more like females (Campbell 1998). And, conversely, males in some anatomically "dioecious" species may occasionally produce seeds and females may produce occasional pollen. Furthermore, the functional or realized gender of a plant can be influenced by many ecological factors such as the effectiveness of pollen vectors, temporal patterns of pollen and seed production in the population, and the degree of self-fertilization (Devlin and Stephenson

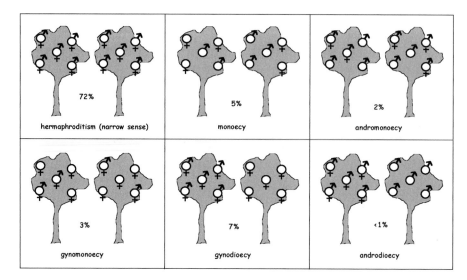

FIGURE 2.2 Pictorial descriptions of six major categories of plant sexual systems that include dual-sex individuals. Each diagrammed plant has five flowers which, depending on the sexual system, can be male, female, or bisexual (as indicated by the symbols). Also shown are the approximate percentages of flowering plant species (angiosperms) that display each of these sexual modes. Additional sexual systems with dual-sex individuals also exist, including various expressions of trioecy (the joint presence within a species of hermaphrodites, males, and females), as well as serial sex changers (species in which a typical individual is functionally either male or female at any one time but may switch sexes within or across breeding seasons).

1987). Operational gender is probably more germane than anatomical gender when considering many genetic and evolutionary issues, including the topic of sex allocation to male versus female functions (Lloyd 1984, 1987a,b; Lloyd and Bawa 1984). Nevertheless, a plant's anatomical gender clearly impacts its sexual function to a considerable extent, both directly (via gamete production) and indirectly (e.g., by influencing pollination patterns).

When defined in functional terms, plant species with dual-sex specimens encompass all of the anatomical systems described in figure 2.2, because in each case at least some individuals can operate as both male and female during a lifetime. Thus, in this chapter, the sexual systems gynodioecy and androdioecy will be included under the umbrella of dual sexuality at the population level, as will monoecy, gynomonoecy, and andromonoecy; and, at the individual level, a hermaphrodite may refer to any dual-sex specimen in any non-dioecious plant species. These usages should facilitate comparisons with the animal sexual systems that will be described in chapters 3 and

4. The many botanical terms in standard use can be confusing but they also help to emphasize a broader conceptual point: sexual systems in plant species actually comprise a biological continuum (Lloyd 1980; Ashman 2006) that is flanked by strict dioecy at one end and by strict dual-sexuality (either monoecy or narrow-sense hermaphroditism) at the other end, and between which lie both quantitative blends and polymorphic combinations of various intermediate dual-sex conditions, either anatomically or functionally or both. Furthermore, it is quite possible that some of the intermediate sexual conditions might themselves be evolutionarily stable (as opposed to merely transitional and ephemeral).

Alternative Sexual Systems: Natural History and Examples

Each of the following sections will describe one or a few species representing each broad category of functional dual sexuality in plants. These sections also will serve to introduce some of the biological, ecological, and evolutionary considerations that will be germane to deliberations (later in this chapter) on the adaptive significance of alternative sexual systems in plants.

Hermaphroditism

In this dual-sex system, each flower on a plant is perfect, i.e., includes functional male and female components. Depending on the species, the arrangements of these sexual parts in time or space may either permit or discourage the union of pollen and eggs from the same flower. When this structural arrangement permits such unions, self-fertilization (selfing) is often the result, whereas when the arrangement of the sexual parts inhibits such unions, outcross fertilizations (involving the union of pollen and eggs from different individuals) may tend to be promoted.

In cleistogamous species, self-fertilization effectively is mandated by a flower configuration in which fertilization events typically occur within a closed bud. About 70 angiosperm species are documented to be almost fully cleistogamous (Culley and Klooster 2007). Darwin (1876, 1877) interpreted cleistogamy in particular, and selfing in general, to be reproductive alternatives of last resort for a plant, arising only when pollinators have become scarce or for some other reason have stopped visiting the flowers. Conversely, selfing is discouraged when male and female parts of a perfect flower mature at different times during a plant's development, a situation known as dichogamy (box 2.1). Selfing may also be discouraged when the male and female flower parts are spatially separated, a situation known as herkogamy (box 2.1), which pertains specifically to any perfect flower in which the male and female components are positioned in such a way as to inhibit self-fertilization.

BOX 2.1   Dichogamy and Herkogamy

Dichogamy is the temporal separation within a plant of the production or dissemination of male and female gametes, and it is one of the most widespread floral mechanisms in angiosperms (Bertin and Newman 1993). Either the pollen is shed before the stigmas are receptive (protandry), or the stigmas are receptive before the pollen is shed (protogyny). These two expressions of dichogamy are quite analogous to the phenomena of protandry and protogyny in various animals that are sequential hermaphrodites (chapters 3 and 4). Functionally, they tend to reduce rates of self-fertilization and increase the rates of outcrossing (Harder et al. 2000).

Historically, many researchers followed Darwin (1862b) in regarding dichogamy as a fitness-enhancing mechanism for reducing the inbreeding costs that otherwise attends selfing. However, a survey of angiosperm species (Bertin 1993) revealed that self-incompatible (SI) plants (see box 2.5) were just as likely as self-compatible (SC) plants to be dichogamous. This led some researchers to propose that the amelioration of stamen-carpel interference (SCI) on pollen transfer to other plants is the primary function of dichogamy (Lloyd and Webb 1986; Barrett 2002b). The SCI phenomenon occurs when the carpel compromises interplant pollen export from the anthers or, conversely, when anthers compromise interplant pollen import to stigmas. Much of the interference is simply mechanical, but some can result from pollen that is used in self-fertilization. SCI can lead to "pollen discounting": a reduction in the amount of pollen available for export and cross-fertilization. Interference appears to be an unavoidable consequence of the close physical proximity of carpels and stamens within a perfect flower, and apparently it can be an important added expense (apart from inbreeding depression *per se*) that can influence the evolution of plant mating systems (Harder and Wilson 1998).

Geitonogamy refers to the pollen transfer between different flowers on the same plant (de Jong et al. 1993), which when successful is also self-fertilization (Lloyd and Schoen 1992). Harder and Barrett (1996) showed that geitonogamy and pollen discounting tend to be greater in plants with larger inflorescences. Thus, in animal-pollinated species, the evolution of flower size probably represents a compromise between the benefits of maximizing floral visitation rates by pollinators (pollinators tend to be attracted to showier flowers) and the costs of geitonogamy and pollen discounting (Holsinger 1996; Snow et al. 1996).

Herkogamy is, in some respects, the spatial analogue of dichogamy. The term refers to any perfect flower whose male and female parts are positioned in such a way as to inhibit self-fertilization (Webb and Lloyd

(continued)

BOX 2.1   (*continued*)

1986; Van Kleunen and Ritland 2004). Thus, many of the ecological and genetic selection-based considerations (such as inbreeding avoidance) that apply to the evolution of dichogamy probably apply to the evolution of herkogamy as well.

In some plant species, perfect flowers on different individuals come in distinct structural morphs, a situation known as heterostyly. In distyly (the most common form of heterostyly), perfect flowers on different individuals display either a pin morph with short stamens and long carpels, or a thrum morph with long stamens and short carpels (fig. 2.3). Because stamens of different length place pollen in different parts of a pollinator's body, pollen grains originating from the long stamens of thrum plants tend to reach mostly the long carpels of pin plants (and *vice versa*), thus promoting cross-pollination between the two morphs. Another type of heterostyly is tristyly, in which three distinct morphs of perfect flower coexist within a species. Heterostyly is widespread in plants, represented in more than two dozen taxonomic families (Givnish 1982). Furthermore, gene-based incompatibility systems routinely operate in conjunction with these anatomical systems of plants to influence the mating system of a species. Specifically, different sexual morphs in heterostylous species tend to be genetically incompatible with individuals of the same morph.

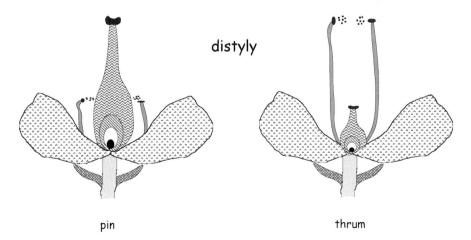

distyly

pin

thrum

FIGURE 2.3   Diagrammatic representation of the most common form of heterostyly in plants.

The flowering rush, *Butomus umbellatus* (Butomaceae), is an emergent aquatic species native to Europe, but widely introduced in North America where it inhabits the margins of lakes and rivers. It is an example of a dichogamous hermaphrodite (Bhardwaj and Eckert 2001), with each flower's male and female functions operative at different times. The common primrose, *Primula vulgaris* (Primulaceae) (fig. 2.4), is an herbaceous perennial native to Europe; it is a distylous hermaphrodite. In 1862, Charles Darwin published a paper describing the dimorphic condition of its pin and thrum flowers, which promotes high rates of outcrossing (Cahalan and Gliddon 1985). Many other *Primula* species are also heterostylous, as are various *Linum* species (Linaceae) including flax or linseed (*L. usitatissimum*), *Lythrum* species (Lythraceae) including purple loosestrife (*L. salicaria*), and many species of *Cryptantha* (Boraginaceae). Considering all flowering plants, hermaphroditism is by far *the* most common anatomical and functional sexual system.

FIGURE 2.4   The common primrose, *Primula vulgaris*, a distylous hermaphrodite.

Monoecy

Pine trees (Pinaceae) and their relatives in the families Taxodiaceae (including sequoias) and Cupressaceae (cypresses and allies) are predominantly monoecious. Givnish (1980) examined 377 of these coniferous plant species in 40 genera, and found that typical individuals in 315 of them (83%) bear both male and female flowers. However, about 85% of 427 other species of gymnosperms in 34 genera proved to be dioecious. Essentially all gymnosperms are wind-pollinated, but Givnish noted strong associations in these naked-seed plants between monoecy and wind-dispersed seeds on the one hand, and between dioecy and animal-dispersed seeds on the other.

Andromonoecy

The horse nettle, *Solanum carolinense* (Solanaceae) (fig. 2.5), is a self-incompatible perennial in which individual plants carry male and bisexual flowers at the distal and basal portions, respectively, of the inflorescences. The male flowers are smaller, have reduced nonfunctional carpels, and do not produce fruit (Solomon 1986; Anderson and Symon 1989). The showy anthers attract pollinators, which are mostly large-bodied bees.

The horse nettle has been the subject of genetic paternity analyses via molecular (allozyme) markers (Elle and Meagher 2000). These studies, conducted on experimental plots, uncovered several otherwise hidden facts, such as: different plants vary in functional gender from completely female to completely male; a plant's reproductive successes as male versus female are negatively correlated (consistent with the notion of reproductive trade-offs between male and female functions); and male reproductive success increases with higher percentages of male flowers on a plant, but not with increases in total flower count. Thus, male and bisexual flowers apparently differ in their contributions to a plant's reproductive success.

Cultivated forms of the melon (*Cucumis melo*) provide another well-known botanical example of andromonoecy (Poole and Grimball 1939). This species is of special interest because it has been the recent subject of detailed molecular dissections revealing how unisexual and bisexual flowers probably were interconverted, evolutionarily (Martin et al. 2009). At least two interacting genes are involved: one that encodes an ethylene biosynthesis enzyme that represses stamen growth in female flowers; and another that indirectly represses the expression of the first locus and thereby promotes stamen development. The authors present a model whereby these two loci mechanistically interact to control the development of the melon's alternative flower types. Such molecular dissections are likely to become more commonplace for other dual-sex plant species as well, as genetic researchers move increasingly into the genomics era.

FIGURE 2.5   The horse nettle, *Solanum carolinense*, a species with andromonoecy.

Gynomonoecy

Perennial herbs in the genus *Aster* (Asteraceae), with about 200 species mostly in North America, provide examples of gynomonoecy. The flower heads on each plant include two floral types: female (ray) flowers and bisexual (disk) flowers. The female flowers typically open first, followed by the bisexual flowers that initially produce pollen and later display receptive stigmas. Thus, each plant in effect shows interfloral protogyny (female-first function) and intrafloral protandry (male-first function).

The adaptive benefits (if any) of gynomonoecy are unclear, but one possibility is that the presence of female as well as hermaphroditic flowers permits great phenotypic flexibility in how individuals allocate resources to male versus female reproductive functions. Depending on ecological circumstances (including gamete pools in the broader population), an individual might profit in terms of genetic fitness by adjusting its portfolio of investments in male versus female gametes. Similar kinds of adaptive explanations are sometimes invoked for systems of monoecy and andromonoecy (Charnov and Bull 1977; Willson 1983). Bertin and Kerwin (1998) tested this phenotypic flexibility hypothesis in greenhouse experiments with various asters, but found no evidence for consistently different patterns of sex allocation in different environmental circumstances. The authors thus favored an alternative hypothesis: that female flowers improve an aster's mean genetic fitness in part by increasing the attractiveness of its flower heads to pollinators.

Gynodioecy

Approximately 7% of flowering plant species consist of females plus hermaphroditic (bisexual) individuals; more than 350 such gynodioecious species from approximately 40 plant taxonomic families have been described (Kaul 1988). Conventional wisdom is that gynodioecy evolves routinely from cosexuality—in part as a selectively driven mechanism for avoiding inbreeding depression—as an intermediate stage in the emergence of dioecy (Charlesworth and Charlesworth 1979; Delph and Wolf 2005). Because pure females can only outcross whereas hermaphrodites perhaps can self-fertilize, at least some females that arise in a partially selfing hermaphroditic species subject to inbreeding depression would probably produce offspring of higher fitness than their hermaphroditic counterparts. In particular biological circumstances, this could set in motion a sequence of evolutionary events that culminates in dioecy (discussed further below). The process might begin when a maternally inherited cytoplasmic mutation causing male sterility (i.e., generating pure females) arises and spreads in a population, thus converting a population from pure hermaphroditism to gynodioecy. In principle, gynodioecy also might arise in some cases via a breakdown of ancestral dioecy; and regardless of origin, it might in some cases be evolutionarily stable (Gouyon and Couvet 1988; Asikainen and Mutikainen 2003).

Sidalcea oregana (Malvaceae)—an herbaceous perennial of mountain meadows in the Sierra Nevada range of western North America—is a gynodioecious species (fig. 2.6) that was the subject of an empirical investigation (Ashman 1994) of reproductive allocation as a function of gender. The goal in that study was to test the postulate that females, relieved of the burden of male function, might reallocate additional resources to female function (compared to hermaphroditic individuals) and thereby compensate for their genetic disadvantage of not being able to produce progeny via pollen. The empirical results suggested that females of *S. oregana* do indeed reallocate some additional resources (that otherwise could have produced pollen) to generate seeds.

FIGURE 2.6   The Oregon checkerbloom, *Sidalcea oregana*, a species with gynodioecy.

Androdioecy

Androdioecy is among the rarest of sexual systems in both plants (about 50 species) and animals (about 36 species) (Pannell 2002; Vassiliadis et al. 2002; Weeks et al. 2006). Mathematical models suggest that this rarity may reflect a restrictive evolutionary condition: males theoretically can be maintained by selection in a population of hermaphrodites only when their fertility is more than double the male fertility of hermaphroditic individuals, who can function as dams as well as sires (Charlesworth 1984). In other words, all else being equal, males in an androdioecious population normally would seem to be at a distinct disadvantage because they have only one genetic pathway to fitness (via pollen) rather than a hermaphrodite's two pathways (via pollen and eggs). Furthermore, androdioecy (unlike gynodioecy) seems unlikely to evolve from hermaphroditism via selection pressures to avoid inbreeding, because selfing in a partially hermaphroditic population would make many eggs unavailable for fertilization by pure males (Lloyd 1975; Charlesworth and Charlesworth 1978). So, when androdioecy does occur, it might be expected to be more prevalent in species that are obligate outcrossers rather than selfers (Eppley and Pannell 2007), and perhaps to have evolved from ancestral taxa that were dioecious rather than cosexual.

One androdioecious plant species is the Durango root, *Datisca glomerata* (Datiscaceae), a perennial flower (Liston et al. 1990) (fig. 2.7). Consistent with the expectations outlined above, this species does indeed have high outcrossing rates (65–92%), as deduced by marker-based studies of genetic parentage (Fritsch and Rieseberg 1992); and its immediate ancestor was dioecious rather than hermaphroditic, as deduced by mapping the sexual systems of *D. glomerata* and related species onto a molecular phylogeny (Rieseberg et al. 1992; Swensen et al. 1998). The PCM approach employed in these latter studies, coupled with genetic analyses of the sex-determination mode, further indicated that the hermaphroditic form arose in an ancestral dioecious population in which a new recessive mutation in the nuclear genome allowed females to produce pollen (Wolf et al. 2001).

Trioecy

Species with this extremely rare breeding system are comprised of separate individuals each bearing only male flowers, or female flowers, or perfect flowers (Sakai and Weller 1999). Trioecy (also termed *subdioecy* by some authors) is documented in two species of cacti (Cactaceae): the beavertail cactus, *Opuntia robusta*, and the giant Mexican columnar cactus, *Pachycereus pringlei* (Fleming et al. 1998) (fig. 2.8). In the latter, populations in Sonora and Baja tend to be trioecious whereas populations elsewhere are gynodioecious, thus suggesting that trioecy probably evolved from gynodioecy via the addition of males at some locales. This columnar cactus thus provides a striking example of intraspecific geographic variation in plant sexual systems.

FIGURE 2.7   The Durango root, *Datisca glomerata*, a species with androdioecy.

Sex-changers

Serial or sequential hermaphroditism is very rare in plants (Freeman et al. 1980a), but a good example is provided by the jack-in-the-pulpit, *Arisaema triphyllum* (Araceae) (fig. 2.9). Populations of this perennial herb of eastern North American woodlands usually consist of distinct male and female individuals, with small adults usually being males and the largest adults invariably being females. The sex is not genetically fixed, however (Policansky 1981; Bierzychudek 1982). Each year, about 50% of the mature individuals change from male to female or *vice versa*. A given specimen might undergo

FIGURE 2.8   The Mexican columnar cactus, *Pachycereus pringlei*, a species with trioecy.

FIGURE 2.9   Jack-in-the-pulpit, *Arisaema triphyllum*, a sequential hermaphrodite.

several such annual sex changes during its lifetime. The association between the sex and the size of a plant suggests that a size threshold might exist below which female reproduction is prohibitively costly (Bierzychudek 1984).

Dioecy

Strict dioecy (the antithesis of dual sexuality within the individual) is widespread and rather common in gymnosperms, notably in the Cephalotaxaceae, Cycadaceae, Ephedraceae, Ginkgoaceae, Gnetaceae, Podocarpaceae, Stangeriaceae, Taxaceae, Welwitschiaceae, and Zamiaceae (Givnish 1980). In angiosperms, dioecy is also taxonomically widespread (Givnish 1982), but only 4–6% of flowering plant species (many of which are tropical forest trees; Bawa and Opler 1975) consist exclusively of separate male and female individuals (Renner and Ricklefs 1995).

Many dioecious plant species have chromosomal modes of sex determination that are "strikingly similar" to the familiar male-heterogametic XY systems of mammals and some other animals (Charlesworth 2002). However, the mere fact that most plant species are not dioecious (but rather include dual-sex individuals, either as monoecious specimens or as hermaphrodites) implies in general that "sex determination" in plants is seldom dictated solely by an individual's genetic constitution *per se*. Instead, an individual's gender in most plant species (including within and among the various branches and flowers of dual-sex individuals) is typically a complex genetic and epigenetic outcome of physiological and developmental processes that activate or repress multiple male-specific or female-specific genes during ontogeny (Irish and Nelson 1989; Dellaporta and Calderon-Urrea 1993; Juarez and Banks 1998).

Cosexuality Versus Dioecy

In many taxa of seed plants (and in sharp contrast to most animal groups; chapters 3 and 4), some form of functional dual sexuality (either monoecy or hermaphroditism) was probably the ancestral population condition whereas dioecy is a derived state. The scientific challenges have been to estimate the number of independent evolutionary transitions from dual sexuality to dioecy, reconstruct the intermediate stages, identify possible evolutionary reversals, dissect the genetic and physiological changes involved, and understand how natural selection often might drive such transitions. Darwin presaged all of these challenges in 1877: "There is much difficulty in understanding why hermaphroditic plants should ever have been rendered dioecious." He also identified two major selective advantages for dioecy that still drive much of current botanical thought: the avoidance of inbreeding, and the reallocation of resources from one sex function to another (Dorken and

Mitchard 2008). With respect to the latter, Darwin (1877) wrote, "if a species were subjected to unfavourable conditions . . . the production of the male and the female elements . . . might prove too great a strain on its powers, and the separation of the sexes would then be highly beneficial."

The problem of deciphering the evolution of alternative sexual systems in plants is tantamount to deciphering the reasons for changes in patterns of sex allocation (Goldman and Willson 1986). This can be seen by noting, for example, that during any transition to dioecy from a condition involving dual-sex individuals, each individual by definition has shifted its reproductive investment strategy from producing two types of gamete (male and female) to producing only one type (male or female). Botanists have focused considerable attention on the ecological and genetic forces that might drive such transitions.

Ecological Considerations

Many researchers have searched for ecological correlates of dioecy that might help to explain the distributions of this relatively uncommon sexual mode in plants (Vamosi et al. 2003). For example, associations have been reported in various flowering plant taxa between dioecy and each of the following: fleshy fruits and animal-dispersed seeds (Givnish 1980; Muenchow 1987; Vamosi et al. 2007), pollination by small insects (Bawa 1980), abiotic pollination by wind (Freeman et al. 1980b) or water (Renner and Ricklefs 1995), specialist mode of pollination with flowers adapted to specific pollinators (Renner and Feil 1993), generalist mode of pollination (Bawa 1994), herbivory on flowers (Cox 1982; Ashman 2002), woodiness and long lifespans (Fox 1985; Sakai et al. 1995a), climbing habit (Renner and Ricklefs 1995), moisture regimes (Conn et al. 1980; Weller et al. 1990; Sakai et al. 1995a), geographic considerations (Baker and Fox 1984), and various other factors (reviewed in Sakai and Weller 1999). Of course, some of these parameters may be non-independent.

Biological rationalizations often are provided for the observed correlations. For example, reported associations of dioecy with inconspicuous flowers (as in wind-pollinated species or those pollinated by small insects) were attributed to the possibility that non-discriminating pollination favors the evolution of female flowers, which otherwise are disadvantageous to plants because they may not attract large-animal pollinators that use pollen as a food source (Charlesworth 1993). A related idea is that females can more readily invade populations that invest little in pollinator attraction, perhaps because such females are freer to allocate more resources to fruit production (Charlesworth and Charlesworth 1987a).

Using a published phylogeny for angiosperms as historical backdrop, Vamosi and colleagues (2003) addressed possible correlates of the dioecy condition with seven ecological and life-history attributes: tropical distribution, woody growth form, abiotic pollination, small inconspicuous flowers and inflorescences, many-flowered inflorescences, and fleshy fruits. The authors' phylogenetic analysis yielded "no general support for the hypothesis that dioecy is more likely to evolve in lineages already possessing the seven attributes we considered." More generally, the many exceptions and even contradictory trends in the various ecological and life-history correlates of dioecy across diverse plant taxa do not inspire confidence that any single parameter or small set of parameters exerts an overriding causal impact on the evolution of plant breeding systems. Instead, an emerging sentiment in botany seems to be that selection pressures from myriad ecological and life-history factors probably interplay with genetic mechanisms and phylogenetic constraints to influence the evolution of breeding systems in plants (Thompson and Brunet 1990; Ashman 2006; Mazer et al. 2007).

Phylogenetic Character Mapping

PCM is used widely in the evolutionary analysis of plant breeding systems (Weller and Sakai 1999; Vamosi et al. 2003). Donoghue (1989) provided an early example of PCM as applied to hermaphroditism versus dioecy in a diverse subset of higher plant taxa. Although researchers had long suspected that the separate-sex condition in angiosperms evolved on multiple occasions from ancestral cosexuality (Lewis 1942; Westergaard 1958; Charlesworth and Charlesworth 1978), Donoghue's study helped to confirm this hypothesis and clarified precisely where in the family-level phylogeny some of the evolutionary transitions probably took place: near the phylogenetic roots of the Myristicaceae, Amborellaceae, and Lactoridaceae, for example (fig. 2.10). Subsequent PCM analyses on finer taxonomic scales (e.g., Soltis et al. 1999) provisionally identified many more such instances, leading to the sentiment that dioecy probably arose on more than 100 separate occasions in flowering plants (Charlesworth 2002). Some of the PCM analyses further suggested that dioecious lineages tend to have rather short evolutionary longevities (Heilbuth 2000).

Donoghue (1989) likewise applied PCM to conifers (fig. 2.10) and thereby documented several independent origins of dioecy from cosexuality, plus two secondary evolutionary reversals to monoecy. Dioecy also proved to be a derived state with multiple origins in extant species representing several basal evolutionary lines of gymnosperm seed plants including cycads, *Gingko*, and *Gnetum* (Donoghue 1989).

Other PCM studies have focused on sexual systems of particular plant taxa and on more specific subcategories of dual sexuality. In perhaps the first such analysis, Hart (1985a,b) demonstrated that dioecy evolved from gynodioecy on at least two separate occasions in the genus *Lepechinia* (Lamiaceae). Likewise in the genus *Silene* (Caryophyllaceae), dioecy arose from gynodioecy at least twice (Desfeux et al. 1996). For *Siparuna* (Siparunaceae), Renner

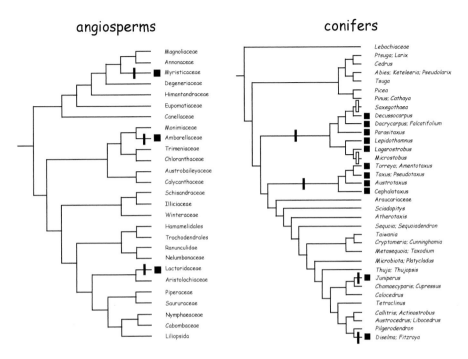

FIGURE 2.10 Phylogenies for angiosperm lineages (left) and conifers (right) showing several independent evolutionary origins (black crossbars) of dioecy (black boxes) from ancestral cosexuality (after Donoghue 1989). In the conifers, two evolutionary reversals (white crossbars) also were deduced.

(1998) deduced that monoecy was the ancestral condition from which dioecy later eventuated. In another such example, Sakai and colleagues (2006; Weller et al. 1995) provisionally reconstructed the course of breeding-system evolution in a spectacular adaptive radiation of Hawaiian endemics in the genus *Schiedia* (Caryophyllaceae). The ancestral condition for the assemblage appears to be hermaphroditism, from which gynodioecy arose at least twice (fig. 2.11): once in *S. apokremnos*, and perhaps again near the base of a large clade (top portion of the figure). In this latter clade, further transitions to dioecy then ensued, plus one clear evolutionary reversion to hermaphroditism (in *S. lydgatei*).

The study by Sakai and coworkers (2006) illustrates several limitations as well as strengths of PCM in reconstructing ancestral phenotypes. Whereas some conclusions seem quite secure (such as the presence of hermaphroditism at the base of the tree, and its reemergence as a derived state in *S. lydgatei*), others remain provisional due to uncertainties regarding: (a) the precise topology of the phylogeny itself (exacerbated in this case by the

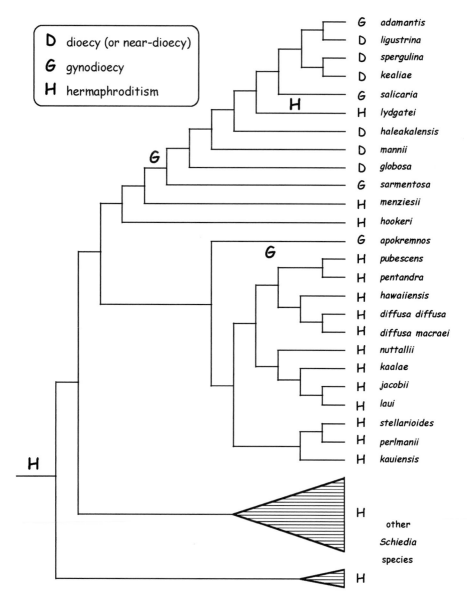

FIGURE 2.11 PCM of alternative breeding systems in 32 species of *Schiedia* (after Sakai et al. 2006). The reconstruction is provisional due in part to uncertainties in the branching order of particular clades (see text).

fact that the adaptive radiation took place in the relatively short time-span of the past 5–7 million years); and (b) ambiguities about the most plausible reconstructions of ancestral states at internal nodes. For example, for the 12 *Schiedia* species at the top of figure 2.11, several alternative scenarios for the number and direction of evolutionary conversions between hermaphroditism and gynodioecy are almost equally parsimonious.

Transitional States

If cosexuality is the ancestral condition in plants and dioecy usually is derived (both overall, and in many subsidiary clades), then the transitional states might represent some mixture of dual-sex and separate-sex individuals, in which case the intermediate evolutionary condition(s) could be androdioecy or gynodioecy (by definition).

The proposed routes to dioecy through gynodioecy and androdioecy are diagrammed in figure 2.12. Under the gynodioecy pathway, the three successive stages are as follows: I, a male-sterile mutant (thus generating females) enters the cosexual population; II, selection acts to decrease allocation to female function in the original hermaphroditic morph (shown in the diagram as a diminished number of eggs in the perfect flower); and III, female function is lost entirely in the pollen-producing morph, transforming the original dual-sex condition into pure male. Thus, the population has transitioned from ancestral cosexuality to a derived condition of dioecy. Alternatively, under the androdioecy pathway, the three stages would be as follows: I, a female-sterile mutant (thus generating males) enters the cosexual population; II, selection acts to decrease allocation to male function in the original dual-sex morph (shown in the diagram as reduced stamens); and III, male function is lost entirely in the seed-producing morph, transforming the original dual-sex condition into pure female. Thus, the population again would arrive at a state of dioecy.

In either evolutionary pathway, the selective pressure that presumably enters the process in stage II probably stems from a form of frequency-dependent selection (Delph 2003) that is quite analogous to the frequency-dependent selection process for sex-ratio evolution in dioecious species (chapter 1). Namely, as unisexual plants increase in frequency in a population of dual-sex individuals, the fitness of functional hermaphrodites might be enhanced if such individuals bias gamete production toward whatever sexual function the unisexuals lack. Thus, this frequency-dependent bias would be toward hermaphrodites performing the male function in the gynodioecy pathway, and toward hermaphrodites performing the female function in the androdioecy pathway.

Either of the two pathways might also terminate before completion of the final step, thus yielding an ostensibly stable condition of gynodioecy or androdioecy. Under ESS theory, however, the conditions for such evolutionary stability appear to be rather restrictive (Charnov et al. 1976; see further discussion below). Intermediate functional or anatomical outcomes might also be facilitated by phenotypic plasticity, because many plant species seem not to be genetically hardwired for maleness versus femaleness (Delph and Wolf 2005). Finally, neither evolutionary pathway is irreversible; for example, hermaphroditism probably evolved secondarily from dioecy (or, more likely, from gynodioecy) in *Schiedia lydgatei* (Sakai et al. 2006). The more standard route in plants, however, generally seems to be from ancestral dual sexuality to descendent dioecy.

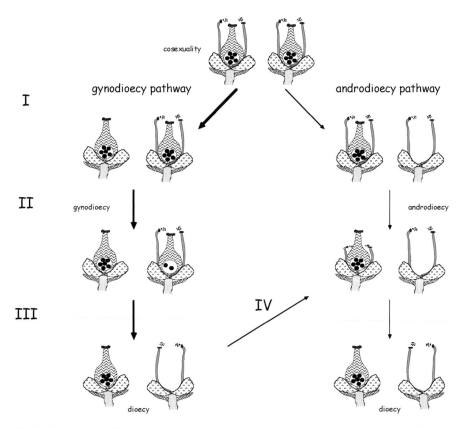

FIGURE 2.12 Candidate evolutionary pathways from cosexuality to dioecy via gynodioecy and androdioecy (after Delph and Wolf 2005). The gynodioecy pathway is more plausible theoretically, and also has been the empirical route employed most often in plants. Also shown is a pathway (stage IV) to androdioecy via dioecy that appears to explain most of the origins of androdioecy in plant species. See text for explanation.

For the two hypothetical evolutionary routes to dioecy pictured in figure 2.12, the gynodioecy pathway is far more common (Webb 1999; Weiblen et al. 2000; Barrett 2002a), probably for at least two reasons. First, male-sterile mutants are more likely than female-sterile mutants to arise and spread in a plant population because of cytoplasmic influences. The cellular cytoplasm and the genomes it contains (cpDNA, mtDNA, and intracellular parasites) are transmitted maternally and thus, from their selfish perspective, are mostly indifferent to the well-being of males. So, maternally inherited male-sterile mutations will tend to spread in a population if the plants that carry them have even a slightly higher seed-set than individuals with non-sterile cytoplasms (as might often be true if such individuals divert to female function some of the energy they otherwise would devote to male function). Female-sterile mutations, by contrast, have no such "Trojan horse" for invading a population consisting solely of cosexual individuals. Cytoplasmic male sterility (CMS) is a common phenomenon not only in plants (box 2.2) but also in invertebrate animals including many arthropods (e.g., Werren 1998).

A second reason why gynodioecy (rather than androdioecy) is the more common evolutionary route from cosexuality to dioecy in plants has to do with inbreeding avoidance. Because self-fertilization in a partially hermaphroditic population would make many eggs unavailable for fertilization by pure males, androdioecy is unlikely to evolve from hermaphroditism via selection pressures to avoid inbreeding (Lloyd 1975; Charlesworth and Charlesworth, 1978). The same cannot be said for gynodioecy, however, because the eggs of any pure females often can be fertilized by pollen from hermaphrodites. So, the fitness penalty that males inevitably experience under androdioecy is not matched by a necessary fitness penalty to females under gynodioecy.

A final point about evolutionary transitions to dioecy via gynodioecy concerns a distinction between two alternative ancestral states of dual sexuality: hermaphroditism and monoecy (Barrett 2002a). From PCM studies, both conditions are strongly suspected to be ancestral to dioecy, depending on the taxon under consideration. For example, as already described, hermaphroditism was probably ancestral to dioecy in the Hawaiian genus *Schiedia* (Sakai et al. 2006); whereas monoecy was immediately ancestral to dioecy in the neotropical shrub family Siparunaceae (Renner and Ricklefs 1995; Renner and Won 2001). If and when monoecy is involved, this sexual system itself normally may have arisen from hermaphroditism, and then secondarily given rise to dioecy either directly or perhaps via gynodioecy (Barrett 2002a). The monoecy route to dioecy probably entails disruptive selection on floral sex ratios leading to greater gender specialization and eventual emergence of separate male and female plants (Barrett 2002a).

If the androdioecy pathway in plants is seldom involved in the evolution of dioecy from cosexuality, how then does androdioecy arise? In most cases, androdioecy has evolved secondarily from dioecy (by hermaphrodites replacing females) rather than directly from cosexuality (by males replacing

BOX 2.2   Cytoplasmic Male Sterility (CMS)

Following its early documentation in more than 140 species of flowering plants, representing 20 taxonomic families (Edwardson 1970; Laser and Lersten 1972), the CMS phenomenon has attracted substantial research attention (e.g., Hanson 1991; Budar et al. 2003). In general, male sterility in various plant species manifests in many guises, including the loss of male organs, meiotic failure, pollen abortion, failure of pollen to dehisce or to germinate, or other processes that in effect cause a loss of male fertility. Although male sterility is sometimes under the control of nuclear genes (see Weller and Sakai 1991), more often it is associated with mutations in mitochondrial (mt) DNA molecules, which are housed in the cellular cytoplasm and show maternal inheritance such that they have no vested interest in reproductive success by males (i.e., via pollen). In principle, any mtDNA gene (such as a male-sterility mutation) that shifts a hermaphrodite's allocation of resources toward the production of ovules rather that pollen would tend to spread in a population. This can set up a powerful conflict of evolutionary interest between nuclear genes (most of which are transmitted through both sexes) and mitochondrial loci. Results of this incessant nuclear-cytoplasmic conflict have been twofold: the origin and spread of CMS in diverse plant taxa; and the frequent evolution of nuclear restorer loci that partially or completely reestablish male fertility (Schnable and Wise 1998; Budar et al. 2003).

CMS arises in two primary contexts. First, in most gynodioecious species that have been studied in genetic detail, male sterility (thus yielding the female condition) is inherited maternally, at least in part (see Sun 1987). By definition, gynodioecious species are composed of hermaphrodites and females but no functional males; thus, operationally, CMS can promote the evolution of gynodioecy from dual sexuality. Furthermore, whenever male function is restored (e.g., by nuclear restorer loci), gynodioecy can revert to some form of dual sexuality. Second, CMS is common in the descendents of crosses between hermaphrodites from different populations or species. Presumably this happens because, in one or both species, a CMS allele had arisen and was countered by a nuclear restorer allele that subsequently swept to fixation (Burt and Trivers 2006). Male sterility then becomes unveiled when the CMS gene and the restorer gene are disassociated via hybridization. Barr (2004) reported what may be an empirical example of such processes in a gynodioecious annual plant native to western North America.

Such intergenomic dynamics imply that CMS is a recurring but often transient condition whose evolutionary comings and goings precipitate back-and-forth shifts between dual sexuality and gynodioecy. They also

(continued)

BOX 2.2   (*continued*)

imply that at least some inter-conversions between these alternative sexual systems may have little to do with the ecological circumstances of particular plant species, and perhaps much more to do with the historical idiosyncrasies of mutations and the evolutionary dynamics of CMS genes and restorer loci *per se*. However, some instances of gynodioecy do seem to be relatively stable as gauged by their evolutionary persistence. For example, gynodioecy is widespread in species of *Plantago* (Plantaginaceae), and *Thymus* (Lamiaceae). Especially in such groups, opportunities might be enhanced for natural selection to adjust parental investments in ovules versus pollen in ways that promote the masculinization of hermaphrodites and the eventual evolution of dioecy from gynodioecy (see the text and fig. 2.12). If so, dioecy itself could be viewed in some respects as an extended evolutionary consequence of CMS genes.

If idiosyncratic mutations are indeed important in shifts between alternative sexual systems in plants, then the mutational scope of mitochondrial versus nuclear genomes becomes of major interest. Presumably, mtDNA has a much smaller mutational spectrum than does the vastly larger nuclear genome. According to Burt and Trivers (2006:181), this may be "a major reason why the nucleus usually wins the conflict with the mitochondria— that is, why most plants are hermaphrodites, not gynodioecious, with male sterility often uncovered only in population or species hybrids."

Finally, another evolutionary factor to consider is the relationship between CMS and inbreeding avoidance. From a mitochondrial perspective, inbreeding is almost always a bad option because a maternally transmitted cytoplasmic gene gains no increased representation in selfed progeny. By contrast, a nuclear gene doubles its representation in selfed versus outcrossed progeny; so, all else being equal, a nuclear gene might wish to avoid selfing only if the fitness of selfed offspring suffers by more than 50% from inbreeding depression. If the avoidance of selfing from the mtDNA point of view is an important function of CMS, then gynodioecy should perhaps be underrepresented in self-incompatible species (Burt and Trivers 2006). Empirically, this may indeed be the case (Charlesworth and Ganders 1979).

hermaphrodites). This seems to be true for *Datisca glomerata* (as already mentioned) and also for *Mercurialis annua* (Euphorbiaceae; Pannell 1997), *Schizopepon bryoniaefolius* (Cucurbitaceae; Akimoto et al. 1999), and *Castilla elastica* (Moraceae; Sakai 2001), among others (Pannell 2002). The only suspected exceptions are *Sagittaria lancifolia* (Alismataceae) and perhaps some species of Oleaceae (Panell 2002). So, in most cases, the evolution of androdioecy in plants probably should be viewed as a stage IV outcome from the left-hand pathway of figure 2.12, rather than as an outcome of stages I or II in the right-hand pathway of that figure.

The discussion in the paragraphs above far from exhausts the extensive speculation and research on how dioecy might evolve from cosexuality in plants. Webb (1999) discussed three other potential evolutionary pathways to dioecy. For example, under one standard version of a "dioecy-from-distyly" model (Rosas and Dominguez 2008), any asymmetrical pollen transfer by pollinators between the distinct floral morphs in a distylous species would tend to favor maleness in the morph with a higher pollen-donation efficiency, and femaleness in the morph that receives more pollen, thus perhaps eventuating in the gradual evolutionary emergence of separate sexes (dioecy) due to the disruptive selection regime initiated by the pollinators. Several authors indeed have demonstrated asymmetrical pollen flow between the floral morphs of distylous species (Stone 1995; García-Robledo 2007). Recently, Rosas and Dominguez (2009) went a step further by documenting the fitness consequences of such asymmetry in a perennial distylous shrub (*Erythroxylum havanense*); their findings appear to be consistent with predictions of the dioecy-from-distyly model.

An Evolutionary Enigma

Because of the phylogenetic directionalities detailed above, the scientific literature on plant sexual systems has focused considerable attention on selective factors that might have promoted the evolutionary transitions from cosexuality to dioecy. However, it should also be remembered that cosexuality in plants is far more common than dioecy. Ergo lies an enigma: the selection pressures for dioecy that have commanded much attention in botany lead away from (rather than toward) the most common sexual system in plants.

Thus, perhaps a better way to approach the topic of alternative sexual systems is to address each mode's potential adaptive costs and benefits, at least some of which are likely to be idiosyncratic to particular evolutionary lineages or ecological settings. Hypotheses about the selective forces that may have shaped the evolution of alternative sexual systems in plants generally fall into two broad categories: those based on population-genetic considerations (mostly inbreeding depression), and those based on ecological considerations. An open mind should also be kept to the distinct possibility that the current distributions of alternative sexual systems in plants may require not only adaptive explanations (from either population genetics or ecology) but also appeals to phylogenetic constraints as well as mutational effects and other nonadaptive intragenomic phenomena (discussed further below).

Selfing Versus Outcrossing

Dual sexuality opens a window of opportunity for self-fertilization that is closed to dioecious species. Most hermaphroditic plants self-fertilize at least occasionally, and many do so routinely (Aide 1986; Goodwillie et al. 2005). In some cases, a selfing event involves male and female gametes within a single perfect flower, a situation known as autogamy. In other cases, it involves separate flowers of an individual, a situation known as geitonogamy. In one early review, Schemske and Lande (1985) reported that about 42% of hermaphroditic plant species display selfing rates greater than 80% (i.e., outcrossing rates of <20%), and many additional species have intermediate rates of selfing and outcrossing in their "mixed-mating" systems (box 2.3).

BOX 2.3   Mixed-mating Systems

A population that engages in both self-fertilization and outcrossing has a mixed-mating system (Clegg 1980; Brown 1989). Much scientific research, using genetic markers, has been devoted to estimating selfing (s) and outcrossing (t) rates (where $s + t = 1.0$) in cosexual plants. A direct approach—multilocus paternity analysis—is applicable when the female parent of each offspring is known (as is often true in seed-bearing plants). Any embryo (inside a seed) that displays alleles other than those carried by its dual-sex mother must have resulted from an outcross event. However, any offspring that displays only the dam's alleles at every gene was probably also sired by that same hermaphroditic individual. A less direct approach—population genetic analysis—can be applied when the dams of progeny are unknown (Hedrick 2000). If a mixed-mating population is at inbreeding equilibrium with respect to s and t, then the observed heterozygosity (H_{obs}) falls below random-mating expectations (H_{exp}). The inbreeding coefficient then becomes $F = (H_{exp} - H_{obs}) / (H_{exp})$, and the estimated selfing rate is $s = 2F / (1 + F)$.

Species that show gynodioecy or androdioecy (or other categories of dual-sexuality) also can have mixed-mating systems. The outcrossing component is guaranteed (assuming that pure females and pure males are reproductively successful), so the behavior of hermaphroditic specimens determines whether selfing (and hence mixed mating) applies as well. Indeed, few if any dual-sex plant populations self-fertilize to the complete exclusion of outcrossing, so mixed-mating systems are quite common in the botanical world (Schemske and Lande 1985; Vogler and Kalisz 2001).

Furthermore, from comparative analyses including PCM, selfing clearly has evolved from outcrossing many times (Stebbins 1970; Wyatt 1988), perhaps on at least 150 occasions in the Onagraceae alone (Raven 1979).

On first inspection, a plant that is capable of selfing would seem to have a selective advantage over an obligate outcrosser. Whereas an outcrossing individual can produce progeny in only two ways (by being either the seed parent or the pollen parent in an outcross event), a self-compatible plant can, in principle, transmit its genes by three hereditary routes: (a) via its own pollen fertilizing its seeds (selfing); (b) via its own seeds after fertilization by foreign pollen (one outcrossing route); and (c) via its own pollen that fertilize foreign seeds (another outcrossing route). The fact that selfing has not fully replaced outcrossing in most plant taxa indicates that additional evolutionary considerations must come into play.

Inbreeding

From a genetic vantage, it does not matter whether a specimen that self-fertilizes belongs to a population that displays monoecy, andromonoecy, gynomonoecy, androdioecy, gynodioecy, or strict hermaphroditism, because in each case two gametes from the same individual unite to form a zygote, which is thus partially inbred. Dioecy, by contrast, ensures outcrossing. One standard hypothesis is that dioecy evolved repeatedly in plants in response to selection pressures for the avoidance of inbreeding depression, which is common and often severe in plants (box 2.4). Consistent with this notion is the widespread occurrence in hermaphroditic plants of outcross-promoting mechanisms. These include the phenomena of dichogamy and herkogamy (see box 2.1), plus various genetic systems that confer self-incompatibility (SI) by blocking fertilization events between pollen and eggs from the same individual (box 2.5).

On the other hand, the fact that selfing is quite common in dual-sex plants suggests that the deleterious consequences of inbreeding are restrained or absent in many cases. One possible explanation entails the phenomenon of genetic purging (box 2.4). If inbreeding depression normally is due to deleterious recessive alleles, then populations with high rates of selfing can become partially cleansed of such alleles by purifying selection against the relevant homozygotes. In other words, the genetic load of harmful recessive alleles can be diminished appreciably in selfing populations that can survive the cleansing process. For such genetically purged populations, the prospects for long-term population viability may then be high. Outcrossing species, by contrast, typically carry a heavy (but nonexposed) genetic load of deleterious recessive mutations mostly in heterozygous condition.

BOX 2.4   Inbreeding Depression in Plants

Darwin (1876) provided descriptions of substantial inbreeding depression in 57 plant species (representing 52 genera in 30 families), and thereby initiated the long-standing view that selfing (an extreme form of inbreeding) is inherently disadvantageous. The inbred plants typically were shorter, weighed less, flowered later, and on average showed a 41% reduction in seed production. Later researchers have confirmed that inbreeding depression is common but also highly variable in magnitude among plant species (Schemske and Lande 1985; Charlesworth and Charlesworth 1987b). Researchers, beginning with Darwin, also have found that the extent of inbreeding depression can vary as a function of environmental conditions (e.g., Dudash 1990; Goodwillie et al. 2005; Pannell 2009).

In many cases, inbreeding depression is due to deleterious recessive alleles that are made homozygous (and their negative fitness effects thereby exposed) following selfing. In such cases, natural selection then can cull many of the harmful alleles from the population, a process known as purging. This purging process is expected to be stronger for alleles of large (rather than small) fitness impacts, so mildly deleterious alleles are likely to escape its effects, especially in small populations where genetic drift can override selection. Purging has been documented in selfing plants as well as in animals (Frankham 1995; Ballou 1997; Husband and Schemske 1997). Although purging is likely to be incomplete (Frankham et al. 2002), in theory it can ameliorate inbreeding depression in species that consistently self-fertilize. Consistent with this notion, a meta-analysis of data sets from the literature revealed that the fitness reductions due to self-fertilization averaged 23% for selfing species and 53% for species that normally outbreed (Husband and Schemske 1996). The analysis also showed that selfing rate and magnitude of inbreeding depression are negatively correlated.

Inbreeding depression has an antipode: outbreeding depression, which is an observed decrease in fitness in crosses between distant relatives (probably due at least in part to a breakup of coadapted combinations of genes). In other words, outbreeding depression can be due to negative interactions between genes that had differentiated in allopatric populations (Schierup and Christiansen 1996). The phenomenon is common in plants, and has given rise to the concept of optimal outcrossing distance: the intermediate geographical scale from which, when plants are collected and artificially crossed, genetic fitnesses are maximized (Waser 1993; Waser and Price 1994). Inbreeding depression and outbreeding depression thus both reside along an evolution continuum of variable fitness effects as a function of genetic relatedness.

BOX 2.5　Genetic Self-incompatibility (SI) Systems

Many species of flowering plants display "self-incompatibility" genetic systems (Franklin-Tong 2008) that help to enforce outcrossing by inducing self-sterility. Each such SI system entails the inability of a fertile hermaphrodite seed plant to produce zygotes after self-pollination (de Nettancourt 1977; Matton et al. 1994; Castric and Vekemans 2004). SI systems in plants are analogous to immune systems in animals in the sense that both entail the genetic capacity of cells to distinguish self from non-self. However, whereas the immune systems in animals evolved to identify and reject non-self entities (notably pathogens), SI systems in plants evolved to identify and reject "self" entities (pollen from the same plant, or from the same type of plant).

The two major types of SI are sporophytic self-incompatibility (Schierup et al. 1998) and gametophytic self-incompatibility (Newbigin et al. 1993). Both of these systems involve genes—known as S-loci—that are highly polymorphic, sometimes harboring 50 or more alleles in the populations that carry them. Much is known about the mechanisms involved, including in some cases the numbers and arrangements of loci and how their biochemical products instigate the blockage of fertilization (de Nettancourt 1977). Much is also known about the evolutionary forces at work. For example, high allelic diversity at SI loci is selectively advantageous because it permits free outcrossing (by lowering the probability that an individual's pollen has allelic matches with other individuals) while simultaneously conferring the avoidance of selfing. The diversifying selection that maintains multiple alleles also tends to buffer against allelic extinction in SI systems, with the evolutionary result that particular genetic polymorphisms at SI loci often persist for long periods of time and can even traverse multiple speciation nodes in evolutionary trees (Ioerger et al. 1990; Dwyer et al. 1991; Savage and Miller 2006).

Gene-based incompatibility systems are widespread in plants, being present for example in more than 50% of angiosperm taxonomic families. They also sometimes operate in conjunction with plant anatomical systems to influence the mating system of a species. For example, gene-controlled SI is typically an integral part of the heterostyly syndrome (see text) and is important to its maintenance; the different sexual morphs of heterostylous species are usually incompatible with individuals of the same morph.

Apart from genetic purging scenarios, researchers have focused most of their attention on two other classes of explanation that might account for the high incidences of selfing in many plant species with dual sexuality: (a) the promotion of coadapted genotypes, and (b) fertilization insurance.

Coadapted Genotypes

Self-fertilization is an intense form of inbreeding, one genetic ramification of which is the buildup of a severe restriction on effective recombination (box 2.6). With continued selfing in a population or lineage, homozygosity rapidly increases (i.e., heterozygosity decays), such that within even a few generations little intra-individual genetic variation remains to be shuffled into new multilocus combinations (see chapter 1). Although meiosis and syngamy continue to operate in selfing populations, they soon become ineffective in generating recombinant genotypes. Furthermore, inbreeding's severe limitation on effective recombination extends across the entire nuclear genome. This contrasts diametrically with the situation in large outcrossing species, where uninhibited genetic recombination in each generation produces vast numbers of novel multilocus genotypes.

The loss of effective recombination can be a two-edged sword for selfing populations. On the one hand, selfing can greatly limit a lineage's genetic scope for response to ecological challenges. Indeed, continued selfing quickly eventuates in individuals that are so inbred as to be, in effect, homozygous clonemates (Avise 2009). On the other hand, for these same reasons continued selfing can help to perpetuate particular multilocus genotypes that may have been favored by natural selection, and that otherwise would be dismantled immediately if the lineage engaged in outcrossing.

Robert Allard and colleagues documented a paradigm empirical example of how self-fertilization in conjunction with natural selection can adaptively shape the multilocus genetic architecture of a plant species. They studied the slender wild oat (*Avena barbata*), a predominantly self-fertilizing species that was introduced to California from its native range in the Mediterranean during the Spanish period about 400 years ago, and then again during the Mission period 250 years ago. In California, the slender wild oat has achieved a remarkable population genetic structure characterized by a great predominance of two coadapted multilocus gametic types (Allard et al. 1972; Clegg and Allard 1972). One genotype, labeled "1,2,2,2,1,B,H" (numbers and letters refer to alleles at each of seven marker genes), is characteristic of semiarid grasslands and oak savannahs bordering the central Sacramento–San Joaquin Valley. The complementary gametic type (2,1,1,1,2,b,h) is more common in moist coastal ranges and in the higher foothills of the Sierra Nevada Mountains. These associations between genotype and environment (notably xeric versus mesic soils) also are maintained over microgeographic scales in transitional ecotones (Hamrick and Allard 1972), notwithstanding the production of recombinant genotypes through occasional outcrossing (at an estimated rate of $t \cong 0.02$; Clegg and Allard 1973).

The authors concluded that genetic variability in *A. barbata* in California has been genomically organized and spatially structured in less than 400 years by intense natural selection operating in conjunction with severe constraints on recombination afforded by the mating system (Allard 1975). These studies added great empirical force to the theoretical argument (box 2.6) that

---

BOX 2.6   Selfing, Recombination, and Genomic Organization

Continued selfing is an intense form of inbreeding that can severely restrict inter-genic recombination, but it can also thereby enhance the opportunity to maintain coadapted gene complexes. This can be illustrated qualitatively as follows. Assume that at each of four bi-allelic loci, the haploid genotypic combinations *a,a,a,a* and *b,b,b,b* (letters refer to alleles at each locus) confer high viability on their diploid bearers and that the other 14 multilocus combinations (*a,b,a,a; b,a,b,a;* etc.) yield less fit individuals. In each generation, favored genotypes will tend to increase in frequency under natural selection, and these will tend to be retained under self-fertilization. By contrast, under random outcrossing, recombination at gametogenesis will tend to undo the effects of selection in maintaining favored multilocus combinations across the generations. The effects on recombination are quite like those of conventional linkage; as shown by Weir and Cockerham (1973), for two loci under a mixed-mating model of selfing and random outcrossing, the expected rate of decline in any nonrandom association of alleles (*D*) is given by:

$$1 - 0.5\{0.5(1+\lambda+s) + [(0.5(1+\lambda+s))^2 - 2s\lambda]^{0.5}\},$$

where *s* is the selfing rate and λ is the tightness of linkage. The important point from this equation is that linkage and selfing tend to retard the loss of disequilibrium in quantitatively identical fashion. For example, for $s = 0.98$ and $\lambda = 0.00$ (or likewise for $s = 0.00$ and $\lambda = 0.98$), the rate of decay from nonrandom allelic combinations is 1.0% per generation.

However, one key difference between the effects of linkage and inbreeding in restricting recombination is that the former acts locally among physically linked loci, whereas the latter acts globally within the genome, restricting recombination between all loci regardless of chromosomal location. Thus, inbreeding in conjunction with selection can in principle serve to organize entire genomes into integrated multilocus systems (Allard 1975).

inbreeding can restrict genetic recombination and thereby enhance the capacity of natural selection to mold coadapted multilocus gene complexes (Jain and Allard 1966; Clegg et al. 1972). At the same time, *A. barbata* "hedges its bets" by retaining the option to outcross, and thereby to generate a plethora of recombinant multilocus genotypes, at least occasionally. A mixed-mating system with predominant selfing can thus promote the maintenance and spread of particular coadapted genotypes that may be successful in the ecological short term, while nevertheless retaining the capacity to produce, via occasional outcrossing, vast genotypic novelty that is probably necessary over the evolutionary longer term.

Fertilization Insurance

Self-fertilizing hermaphrodites have another advantage that obligate out-crossers (including males and females in dioecious species) lack: the capacity to reproduce without a mating partner. Herbert Baker, a well-known botanist, saw this as a key factor underlying the success of self-fertilizing species. Citing his own research on plants (Baker 1948, 1953) and Longhurst's (1955) findings for marine invertebrates, Baker (1955, 1965) concluded that the capacities for self-fertilization and for long-distance dispersal are positively correlated across species, and that a plausible explanation involves the reproductive assurance (fertilization insurance) that comes with being a selfing hermaphrodite. In other words, Baker suggested that natural selection favors selfing capabilities in dispersive or weedy species because even a single individual can be a successful colonist. Plants that are effective colonizers usually have highly dispersive propagules (seeds) that may be carried for long distances by wind, water, or animals. Upon arrival at a distant location, an immigrant's ability to reproduce self-sufficiently (rather than requiring a partner) is often advantageous if not crucial. The empirical association between self-fertilization and colonizing potential is now known as Baker's rule. Pannell and Barrett (1998) have modeled Baker's rule and shown how the concept also applies in principle to plant metapopulations in which both colonization and extinction are frequent.

Consistent with Baker's rule, island floras often seem especially prone to selfing. For example, a disproportionate percentage of the cosexual species endemic to the Juan Fernandez Islands (far off the coast of Chile) are self-compatible (Anderson et al. 2001), as are native floras in New Zealand (Webb and Kelly 1993), the Galapagos Islands (McMullen 1987, 1990), and the British Isles (Price and Jain 1981; but see Williamson and Fitter 1996). Similarly, for 17 invasive plant species surveyed in South Africa, Rambuda and Johnson (2004) found that all were capable of uniparental reproduction (either via selfing or apomixis). This latter study included 13 woody species, suggesting that Baker's rule may apply not only to herbaceous plants (its traditional domain) but also to some shrubs and trees. (Conventional opinion had been that fertilization insurance is unimportant in long-lived perennials because each plant has multiple flowering episodes; Lloyd and Schoen 1992; Bond 1994.) Several authors have touted the general merit of Baker's rule in various other floras as well (Schueller 2004; Busch 2005; Flinn 2006). On the other hand, rules in biology seem made to be broken, and Baker's rule has many apparent infractions or even gross violations in particular settings (Carr et al. 1986; Sun and Ritland 1998; Brennan et al. 2005, 2006). For example, dioecy is unexpectedly common in the extant flora of the Hawaiian Islands, apparently due to the combination of a relatively high rate of colonization by dioecious (rather than hermaphroditic) ancestors plus the repeated evolution of dioecy *in situ* (Sakai et al. 1995b). For plants generally, at least one entirely different hypothesis for the present-day distribution of selfing versus outcrossing taxa also has been raised (Miller and Venable 2000; Miller et al. 2008); but this hypothesis (which involves diploidy versus polyploidy) has not yet been thoroughly evaluated.

In theory, selfing could be advantageous not only in colonizing species but also whenever individuals experience special difficulties in encountering potential mates or in accessing gametes from the opposite sex. For example, in any hermaphroditic species in which population densities are low or individuals are either sedentary (as in most plants) or solitary, outcrossing opportunities could be severely limited and large genetic fitness premiums might therefore attend any capacity to self-fertilize.

Sex Allocation

Another factor clearly impacted by the degree of selfing versus outcrossing in simultaneous hermaphrodites is the optimal allocation of resources to male versus female functions within a plant. At one end of the theoretical continuum are obligate selfers; for any such individual, the relevant gametic pool is within the individual itself. Thus, an obligate selfer normally would do best to invest disproportionately in ova because even a relatively trivial investment in inexpensive male gametophytes should suffice to fertilize all of its eggs (Charnov 1979, 1987a; de Jong et al. 1999). At the other end of the theoretical continuum are obligate outcrossers with random mating; for any such individual, the relevant gametic pool is that of the local deme, i.e., the population well within the range of pollen dispersal. In such situations, the relative fitness payoffs for investing in male versus female gametes should be dictated by the ratio of available male to female gametes in the deme, such that at equilibrium an individual might fare best by investing equally in male and female function. More generally, across the full spectrum of potential selfing rates, sex allocation theory (SAT) predicts an inverse relationship between selfing rate and the relative allocation to male function (fig. 2.13).

A relationship between high selfing rate and diminished allocation to male function is a strong prediction of SAT that has proved to be highly consistent with empirical data (Charlesworth and Morgan 1991, and references therein). Another empirical finding highly consistent with SAT is that a plant's allocation to attractive flowers is typically low in selfers; most selfing species have small flower sizes (e.g., Cruden and Lyon 1985; Ritland and Ritland 1989). (An understandable exception is when self-fertilization events require pollinator visits, as normally is true under geitonogamy [Lloyd 1987b].)

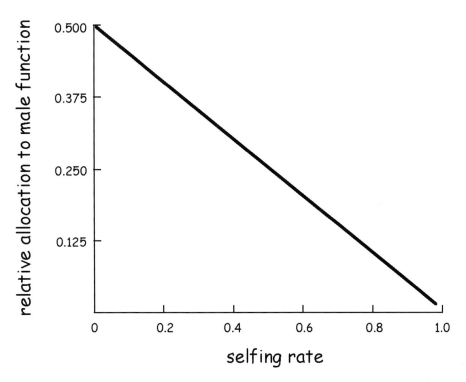

FIGURE 2.13 For simultaneous hermaphrodites, sex allocation theory predicts an inverse relationship between selfing rate and an individual's resource allocation to male (as opposed to female) function (after Charnov 1982). The exact relationship also varies somewhat according to the magnitude of inbreeding depression, d (the thick dark line shown in the figure assumes $d = 0.50$).

However, several other predictions of SAT have proven more difficult to verify observationally or experimentally (Brunet 1992; Campbell 2000). Part of the difficulty may simply be the empirical challenge of measuring relevant parameters such as a plant's lifelong fitness or its resource allocations to male versus female functions (Campbell 1998). Another part of the difficulty, however, may be that some of the basic assumptions of SAT are violated, at least occasionally (e.g., Seger and Eckhart 1996). For example, it may not always be the case that an individual's allocations to male and female reproduction constitute a zero-sums game such that trade-offs between the two inevitably ensue (Campbell 2000; Mazer et al. 2007).

Sexual Selection in Plants

In his pioneering treatise on sexual selection, Darwin (1871) did not exclude the possibility of sexual selection in plants (Willson 1990), but he did focus mostly on animals when developing and illustrating the concept. Nevertheless, botanists now appreciate that the phenomenon of sexual selection (selective pressures arising from competition for mates) is ubiquitous and multifaceted in plants, as it is in animals. A key contributor to this realization was Mary Willson (1994), who adduced multiple lines of evidence for sexual selection in plants and considered the many ramifications of mate competition for plant mating systems and flower phenotypes. Sexual selection in plants can manifest itself in many ways. For example, male flowers are often larger than female flowers, presumably thereby increasing pollinator visits and pollen export, both of which can be interpreted as selective responses to male-male competition for fertilization success. In animals, males routinely compete at several stages of the mating process including mate attraction, sperm delivery to the female, postdelivery sperm competition within the female reproductive tract, and post-fertilization phenomena such as mate guarding. Analogously, male plants routinely attract pollinators (the intermediaries in their "matings" with females), their pollen compete for fertilizations after arriving on a female flower (e.g., Emms et al. 1996), and the flowers or ova that their pollen take out of circulation via successful syngamy register, in effect, genetic victories over competing males. Sexual selection among male plants can extend even to the post-fertilization phase inside a female when, for example, a maternal parent discriminates among embryos sired by different fathers (reviewed in Willson and Burley 1983; Marshall and Folsom 1991), or when zygotes with different sires compete for finite seed space or other maternally provided resources (Cruzan 1990). In short, although sexual selection was not emphasized in the scientific literature on plant mating systems until rather recently, there are good reasons to think that its evolutionary impacts on plant species are as pervasive as they are on animals.

Another important point is that sexual selection likewise can operate in such a way as to impact the evolution of floral and other secondary sexual phenotypes in plant species that include dual-sex individuals. Thus, as will be elaborated in chapters 3 and 4, sexual selection applies with full force to hermaphroditic as well as to separate-sex species. Indeed, the important body of sex allocation theory that arose in the 1970s (Lloyd and Webb 1977; Charnov 1979, 1982) initially focused on the relative fitness contributions of separate male and female functions in hermaphrodites.

SUMMARY

1. In the botanical literature, hermaphroditism traditionally has been reserved for situations in which a plant bears both male and female structures within each bisexual flower. However, in a broader functional sense, the phenomenon of dual sexuality also exists within any species in which at least some individuals produce both male and female gametes. Thus, the topic of dual sexuality in plants also necessitates a consideration of species that display: monoecy (with mixtures of both types of unisex flowers on each individual); andromonoecy (mixtures of bisexual and male flowers on an individual plant); gynomonoecy (mixtures of bisexual and female flowers on a plant); androdioecy (some individuals with bisexual flowers and others with male flowers only); gynodioecy (some individuals with bisexual flowers and others with female flowers only); trioecy (some individuals with bisexual flowers, others with male flowers, and others with female flowers); and serial or sequential sex change.

2. Complicating matters further is the fact that in various expressions of dual sexuality in plant species, the relative proportions of male flowers, female flowers, and bisexual flowers can vary from near-zero to near-100%, both within individuals and within or among populations. Such variation in investment strategies in male versus female gametes also highlights the idea that the topic of sex allocation (SA) is intimately connected to questions about how alternative sexual systems in plants evolve and perhaps are maintained or promoted by various ecological and/or genetical selective pressures.

3. Approximately 95% of extant species of flowering plants (angiosperms) display one form or another of dual sexuality, with hermaphroditism being the most common mode (72%). Dual-sex individuals are also common and taxonomically widespread in gymnosperms (plants bearing naked seeds). Thus, in plants, one form or another of dual sexuality greatly predominates over dioecy (separate sexes) and is usually assumed to be the ancestral condition from which dioecy evolved on many separate occasions. For various plant taxa, the numbers and placements of these evolutionary transitions

to dioecy (and, sometimes, apparent reversals to dual sexuality) have been estimated by phylogenetic character mapping (PCM).

4. From PCM and other evidence, most of the evolutionary transitions to dioecy from cosexuality (either monoecy or strict-sense hermaphroditism, depending on the taxon) seem to have occurred along an evolutionary pathway that entails gynodioecy as an intermediate stage, although various other pathways are known or suspected as well. In most cases examined, androdioecy appears to have evolved secondarily from dioecy, rather than directly from cosexual ancestors.

5. Many studies have addressed the possible ecological correlates of dioecy in plants, but results generally have been inconclusive and sometimes inconsistent across taxa. An emerging sentiment seems to be that myriad ecological factors probably interact with genetic mechanisms and phylogenetic constraints to influence the evolution of alternative breeding systems in plants.

6. Traditionally, dioecy has been interpreted as an outcross-enforcing mechanism that provides an evolutionary escape from the perils of inbreeding depression sometimes otherwise associated with self-fertilization by dual-sex individuals. Many species with dual-sex individuals also appear to have evolved various inbreeding-avoidance mechanisms, including various genetic systems of self-incompatibility, as well as the phenomena of dichogamy and herkogamy (the temporal or positional separation, respectively, of male and female sexual components within a perfect flower).

7. Nevertheless, many species retain self-fertilization capabilities, usually as part of a mixed-mating system that also includes outcrossing (at a collective diversity of rates). At least in theory, populations with high selfing rates sometimes can purge themselves of deleterious recessive alleles and thereby clear one major hurdle of inbreeding depression. Other standard hypotheses for the persistence of selfing in particular taxa include: the possibility for evolving coadapted multilocus genotypes (a prototype example involves the slender wild oat, *Avena barbata*); and the fertilization insurance (reproductive assurance) that selfing provides. The empirical association of selfing with weedy habit and colonization potential in plants is known as Baker's rule, which also can apply to any dual-sex species in which encounter rates for male and female gametes from different individuals are low.

8. Sexual selection theory, which deals with the evolutionary consequences of intraspecific competition for mates, was initially developed and applied mostly to animal systems, but it similarly applies with full force to many plants, including species with dual-sex individuals. Sexual selection in plants can operate both pre- and post-zygotically, and it has had important impacts on the evolution of floral traits and other secondary sexual phenotypes.

CHAPTER THREE

Dual-sex Invertebrates

An estimated 65,000 species (approximately 6%) of known invertebrate animals are hermaphroditic (Jarne and Auld 2006). If we exclude the species-rich insects (none of which appears to be hermaphroditic), the incidence of hermaphroditism in the remaining invertebrate species rises to about 30%. Hermaphroditism is also taxonomically widespread, occurring in 22 of the 32 invertebrate phyla (69%) and being ubiquitous in 12 of them (table 3.1). At the next lower taxonomic level, hermaphroditism is represented in nearly 50% of 85 invertebrate classes (Eppley and Jesson 2008). Hermaphroditism is also common in fungi (which are related more closely to animals than to plants, but will not be discussed here).

Darwin was well aware that the phenomenon of dual sexuality is not confined to plants. Indeed, he published monographs on two invertebrate groups—barnacles (Darwin 1851, 1854) and earthworms (Darwin 1881)—that are primarily or entirely hermaphroditic. He contemplated the evolutionary significance of hermaphroditism in *On the Origin of Species* (1859), and he mused about a rather peculiar observation (given that each hermaphroditic individual houses both male and female organs): "there is reason to believe that with all hermaphrodites two individuals, either occasionally or habitually, concur for the reproduction of their kind"; and "no organic being fertilizes itself for a perpetuity of generations; but that a cross with another individual is occasionally—perhaps at long intervals of time—indispensable." Darwin call this generality a "law of nature," and it still appears to hold; genetic and other data have confirmed that most if not all hermaphroditic lineages (in both plants and animals) outcross at least occasionally, and often routinely. Outcrossing can be important, as Darwin

TABLE 3.1   Incidence of Hermaphroditism in Invertebrate Animals (after Michiels 1998; Jarne and Auld 2006)

Phylum or Group	Common Name	N*	Hermaphroditism
Acanthocephala	thorny-headed worms	1,200	absent
Cycliophora	pandoras	1	absent
Kinorhyncha	mud dragons	150	absent
Loricifera	corselet bearers	9	absent
Myxozoa	slime parasites	1,200	absent
Nematomorpha	horsehair worms	320	absent
Priapula	cactus worms	16	absent
Hemichordata	hemichordates	85	absent
Onychophora	velvet worms	80	absent
Rotifera	rotifers	1,800	absent
Arthropoda	insects, crustaceans, and allies	960,000	rare
Brachiopoda	lamp shells	300	rare
Echinodermata	starfish and allies	6,000	rare
Nemertina	ribbon worms	900	rare
Sipuncula	peanut worms	300	rare
Tardigrada	water bears	600	rare
Nematoda	roundworms	20,000	present
Phoronida	horseshoe worms	15	present
Annelida	segmented worms	14,400	common[1]
Mollusca	mollusks	118,000	common[1]
calcareous sponges	sponges	1,000	ubiquitous
Chaetognatha	arrow worms	100	ubiquitous
Cnidaria	anemones and corals	9,000	ubiquitous
Ctenophora	comb jellies, sea walnuts	100	ubiquitous
demosponges	sponges	8,000	ubiquitous
Ectoprocta	bryozoans	4,500	ubiquitous
Entoprocta	lophophorates	150	ubiquitous
Gastrotricha	gastrotrichs	430	ubiquitous
hexactinellids	sponges	1,000	ubiquitous
Placozoa	sponges	1	ubiquitous
Platyhelminthes	tapeworms, flatworms	13,800	ubiquitous
Urochordata	sea squirts, salps, and allies	1,300	ubiquitous

*N is the approximate number of species in each phylum or group.
[1]But highly variable across different subtaxa (for mollusks, see Heller 1993, and Baur 1998).

(1876) appreciated, in part because of diminished fitness under the inbreeding depression that self-fertilization (selfing) often entails.

Biological Examples

Hermaphroditism comes in many guises in invertebrate animals. What follows are representative examples of hermaphroditic species that are of special interest because of their intriguing natural histories.

Reef-building Corals

As emphasized by George Williams (1975) in his famous strawberry-coral model, corals bear many ecological similarities to plants. Corals are sessile organisms, modular and often branched, in which the polyp is the fundamental building block that is duplicated hundreds of times to form a colony (fig. 3.1). Each polyp in principle is self-sufficient, but in reality each colony tends to be synchronized in physiological features such as the onset of sexual maturity and the timing of gametogenesis. Corals, like plants, can propagate clonally from "cuttings" (Neigel and Avise 1983), but both of these kinds of organisms also reproduce sexually. Most coral species (more than 67%) are simultaneous hermaphrodites; male and female gametes typically are produced inside each polyp (Harrison and Wallace 1991). Thus, by analogy to plants, selfing in corals could entail syngamy of gametes from a single polyp (autogamy) or the syngamy of gametes from different polyps within a colony (geitonogamy). Another analogy is that many corals are broadcast spawners, releasing male gametes into the water much as wind-pollinated plants release pollen into the air. In many coral species, unfertilized ova also are released externally, but in brooding species the ova are retained within the colony and fertilized by sperm that were broadcast into the sea by the same or another colony. For various sessile marine invertebrates, all four combinations of the two sexual modes (gonochoristic versus hermaphroditic) and the two dispersal patterns (broadcast spawning versus brooding) are known (Carlon 1999; Bishop and Pemberton 2006).

Love-dart Snails, Earthworms, and Sea Slugs

Many terrestrial snails (superorder Pulmonata; phylum Mollusca) consist of dual-sex individuals that pair with one another and exchange gametes during outcross mating events. Each hermaphrodite typically functions as male *and* female during each encounter, both donating sperm to and receiving sperm from its partner, via penises and intromission. The received sperm

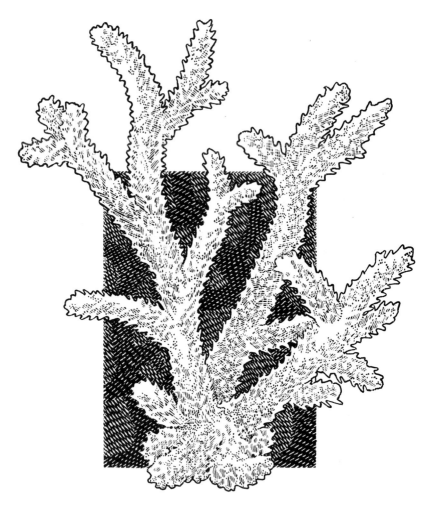

FIGURE 3.1   The staghorn coral (*Acropora cervicornis*), a prominent branching coral on Caribbean reefs.

may either be digested (at rates as high as 99%), or stored for up to two years in special organs known as spermathecal sacs (Rogers and Chase 2001) for later use in fertilizing some or all of the many eggs in a clutch. The reciprocal exchange of gametes at copulation can generate potential conflicts of interest between hermaphroditic partners (Leonard 1990) and thereby precipitate evolutionary "arms races" between male and female sexual functions (Adamo and Chase 1996; Koene and Schulenburg 2005; but see also Leonard 1992, and Chase and Vaga 2006).

The brown garden snail *Cantareus aspersus* (formerly *Helix aspersa*) (fig. 3.2) provides a fascinating example of one such arms race. In this species and in numerous other terrestrial snails, couples shoot "love-darts" at one another in a pre-mating courtship (Adamo and Chase 1988). Each dart is a calcareous spear, forcefully ejected by the thrower, which when well-aimed punctures the recipient's body and delivers a mucus substance. Experiments have indicated that the mucus contains an allohormone (Koene and Ter Maat 2002) that decreases sperm digestion and thereby significantly increases fertilization rate by the successful dart-thrower (Koene and Chase 1998; Chase and Blanchard 2006). This species also displays sperm competition (box 3.1) and early-male sperm precedence (Evannno et al. 2005; see also Baur 1994), meaning that if an individual has multiple mates, sperm from the earliest mate tends to have a fertilization advantage. Thus, a snail can optimize its reproductive success by shooting its dart accurately and mating with a virgin. Over evolutionary time, an apparent arms race between male and female functions has yielded a diversity of shapes and sizes of snail darts (fig. 3.3). It has probably led also to increasingly sophisticated responses in the female reproductive tract (such as the evolution of different compartments for

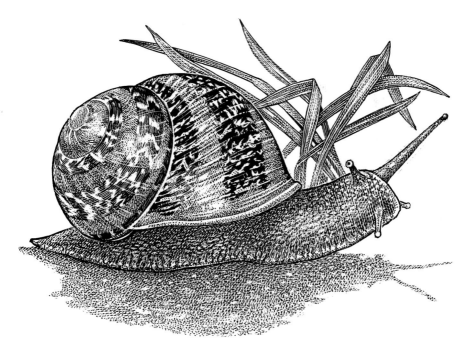

FIGURE 3.2   The brown garden snail (*Cantareus aspersus*), a love-dart species that illustrates simultaneous hermaphroditism.

BOX 3.1   Sperm Competition and Cryptic Female Choice

In many if not most invertebrate species, sperm from two or more males are often placed in direct competition for fertilization of eggs during a female's reproductive cycle. Many morphological characteristics and reproductive behaviors of males have been interpreted as adaptations to meet the genetic challenges resulting from this competition with another male's sperm (Parker 1970). For example, in many worms, insects, and spiders, a male secretes a plug that serves temporarily as a "chastity belt" to block a female's reproductive tract from subsequent inseminations. In many damselflies and dragonflies (Odonata), males have a recurved penis that during mating physically scoops out old sperm (from other males) from a female's reproductive tract, thus helping to account for the tendency of last-mating males to sire disproportionate numbers of progeny (Cooper et al. 1996; Hooper and Siva-Jothy 1996). Other widespread male behaviors that have been interpreted as providing paternity assurance in the face of potential sperm competition include prolonged copulation (up to one week in some butterflies), multiple copulations with the same female, and both pre- or post-copulatory mate-guarding (Parker 1984; Jormalainen 1998).

From a female's perspective, mechanisms to prevent competition among sperm from different males are not necessarily desirable, and this can lead to intersexual conflicts of interest (Knowlton and Greenwell 1984; Eberhard 1998). Furthermore, growing evidence suggests that the reproductive tracts of females often play more active roles than previously supposed in post-copulatory choice of fertilizing sperm (Birkhead and Møller 1993a; Mack et al. 2003). In other words, there is the potential for "cryptic female choice" of sperm. These and related topics have made sperm competition, cryptic female choice, and intersexual conflict among the hottest topics in molecular ecology and evolution during the last quarter century (Smith 1984; Baker and Bellis 1995; Birkhead and Møller 1998).

Sperm competition certainly is not confined to gonochoristic species; hermaphrodites often display the phenomenon as well. Indeed, sperm competition is almost inevitable in any sexual species in which eggs potentially are exposed to sperm from multiple individuals.

---

storing sperm) to the challenges posed by a partner's attempted manipulations of female sperm usage (Koene and Schulenburg 2005).

Earthworms (in the phylum Annelida) likewise are simultaneous hermaphrodites that outcross in paired copulations. Analogous to dart shooting in land snails, some earthworms have specialized setae (bristle-like

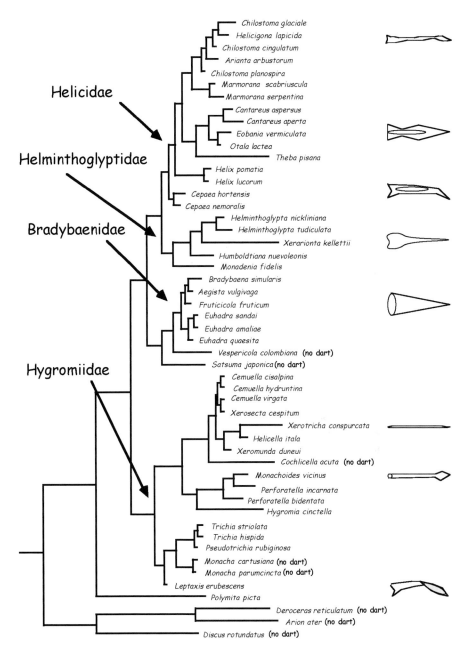

FIGURE 3.3 Phylogeny of land snails and their love-darts (after Koene and Schulenburg 2005). Darts are diagrammed for representative species, but dart sizes and shapes vary considerably among all of the love-dart species (for comparability, all darts are drawn to the same size here). The phylogenetic appraisal was based on nucleotide sequences at a rRNA gene.

structures) that pierce a partner's skin during courtship and inject an allo-hormone inside the body (Koene et al. 2002). Each earthworm, *Lugubris terrestris* (fig. 3.4), has 40–44 such setae, and the substance it injects has been shown to facilitate sperm uptake by the mating partner, thereby stacking the odds in favor of donor success in the fertilization events (Koene et al. 2005).

A more direct route to fertilization is taken by some simultaneously hermaphroditic sea slugs (superorder Opisthobranchia; Mollusca), as well as some leeches (Annelida) and flatworms (Platyhelminthes), which display "hypodermic insemination." In such species (an example being the sea slug *Elysia crispata*; fig. 3.5), each individual has an extendable penis with a sharp style that pierces the partner's skin during copulation and injects sperm into the mate's body cavity (Angeloni 2003). It is unclear exactly how the sperm then travels to fertilize the eggs, but the journey may not be too

FIGURE 3.4   The earthworm *Lugubris terrestris*, another simultaneous hermaphrodite.

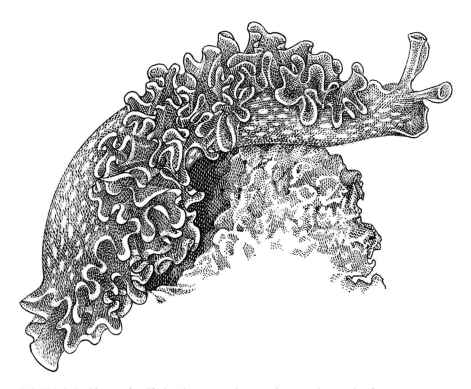

FIGURE 3.5   The sea slug *Elysia crispata*, another simultaneous hermaphrodite.

difficult because an ovotestis (which produces both eggs and sperm) occupies most of the animal's body (Evans 1953).

Sperm-storing Freshwater Snails

Experimental and observational studies of mating systems have been conducted on more than 50 species of freshwater snails (in the superorder Pulmonata), most of which have proved to be simultaneous hermaphrodites with mixed-mating systems of predominant outcrossing and occasional selfing (reviewed in Jarne et al. 1993). The internal reproductive anatomy of these snails is complicated and includes hermaphroditic parts that produce male and female gametes in an ovotestis, female parts including an oviduct that channels eggs to the outside of the snail's body and that channels allosperm (sperm from other males) to the inside, and male parts including a spermiduct through which autosperm (sperm from the focal individual) is evacuated during copulation via a penis. Each outcross event is a unilateral copulation, with one individual playing the role of male and another individual playing the role of female. After the first copulation, a reversal of sex roles by the two participants often occurs, as well as additional copulations either as male or female with other individuals. Many species also have aphallic individuals that lack a penis; these individuals are unable to copulate as a male, but they can copulate as a female and also self-fertilize. Virgins often self-fertilize before their first mating, but thereafter cross-fertilize primarily (unless isolated for long periods from other individuals).

Following a copulation, viable allosperm can be stored within the female parts and used to fertilize eggs for up to several months, after which the allosperm gradually die or become exhausted. This protracted sperm storage (box 3.2) permits cross-fertilizations to continue long after the most recent copulation.

BOX 3.2   Sperm Storage by Females

In many vertebrate and invertebrate species, a female's reproductive tract is physiologically capable of storing viable sperm for a considerable period of time after a copulation event (Howarth 1974; Smith 1984; Birkhead and Møller 1993b; Avise 2004). The duration of female sperm storage varies from a few days in many mammals, to weeks in many insects and birds, to months in some snails, fish, salamanders, and, most remarkably, up to several years in some snakes and turtles (including documented cases in natural as well as captive settings; Pearse and Avise 2001; Pearse et al. 2001). Traditional evidence for sperm storage came from direct observations of live sperm stored in special female storage organs (spermathecae or spermathecal sacs), or from the fact that females long-isolated from males (e.g., in zoos) sometimes continue to produce offspring. For the latter observation, an alternative possibility—that the progeny arose asexually, via parthenogenesis—can be eliminated by molecular parentage analyses; parthenogenetic offspring should be genetically identical to one another and to their mother, whereas progeny produced by sperm storage should be genetically diverse.

For many species, long-term female sperm storage is likely to expand greatly the opportunities for sperm competition and cryptic female choice (box 3.1). The phenomenon applies not only to many gonochoristic species but also to many hermaphroditic species that outcross.

Sex-changing Limpets, Isopods, and Polychaetes

Most hermaphroditic invertebrates express both sexes simultaneously, but sequential hermaphroditism is well represented also, occurring in at least 15 taxonomic classes (Eppley and Jesson 2008). Limpets are marine snails that exemplify protandrous hermaphroditism (Collin 2006); each individual typically begins life as a male but later may switch to female. Individuals are solitary in some limpet species, whereas in others they occur in groups, often in a stacked configuration (Hoagland 1978; Warner et al. 1996). Slipper limpets (genus *Crepidula*) illustrate a typical pattern in the stacking species, which have pelagic larvae that settle onto hard substrates. Within each stack (fig. 3.6), basal individuals (who also tend to be the oldest and largest) are female, whereas distal individuals (who have joined the stack more recently, and thus are smaller) tend to be male. Copulation may occur between any male and female within the stack. If a larva lands on an established stack, it may remain a male for many years, until the stack grows or perhaps until some older individuals die and the focal specimen then becomes a basal-position female. The sex change itself takes about two months, during which time the penis regresses (individuals become aphallic) and female organs develop.

Among invertebrates, sequential hermaphroditism usually takes the form of protandry (male first, as in limpets), but examples of protogyny (female first) also are known (Allsop and West 2004). For example, in a survey of 63 sex-changing species in the subphylum Crustacea (crabs, shrimps, isopods,

FIGURE 3.6   A stack of slipper limpets (*Crepidula fornicata*), a species that illustrates protandrous hermaphroditism.

and allies), 11 species (17%) were protogynous (Brook et al. 1994). Indeed, protogyny is the norm in several isopod taxa. In these species, in which males guard their mates, older and larger males probably have a reproductive advantage over smaller males by virtue of an enhanced ability to thwart cuckoldry by competitors (Brook et al. 1994; Abe and Fukuhara 1996).

In a few invertebrate species, individuals may switch sex repeatedly during a lifetime, in a fashion somewhat similar to the jack-in-the pulpit plants discussed in chapter 2. A case in point is the protandric polychaete *Ophryotrocha puerilis* (phylum Annelida), in which both members of a reproductive pair tend to change sex simultaneously after several spawning events (Berglund 1986). The adaptive significance of this serial sex change may be related to the fact that females grow more slowly than males due to higher female reproductive costs, yet larger females (unlike larger males) are more fecund. In this species, the relative sizes of individuals in a mated pair can change repeatedly over time, and this can create a benefit for repetitive sex changers due to the trade-off between present and future reproductive rates (Berglund 1986).

Androdioecious Clam Shrimps and Gynodioecious Sea Anemones

As in plants, some invertebrate species consist of mixtures of hermaphroditic and single-sex individuals. For example, several species of clam shrimp (class Branchiopoda) are androdioecious, consisting of mixtures of hermaphrodites and males. In these tiny freshwater crustaceans, hermaphrodites lack clasping appendages for mating and thus can only self-fertilize or mate with males (Sassaman and Weeks 1993; Weeks et al. 2005a). Furthermore, based on phylogenetic and biogeographic considerations, androdioecy appears to have persisted in clam shrimps for at least several tens of millions of years (Weeks et al. 2005b). Altogether, 38 invertebrate animal species alive today are documented to be androdioecious (Weeks et al. 2006). Most of these are crustaceans, including (as first described by Darwin in 1851) more than a dozen species of barnacles (class Cirripedia, or Thecostraca), as well as some tadpole shrimp (order Notostraca) (García-Velazco et al. 2009). Androdioecy also is known in a few nematode species (phylum Nematoda), including *Caenorhabditis elegans*, which has been a model organism for scientific research in genetics and developmental biology (Riddle et al. 1997). In *C. elegans* (as in clam shrimps), males are the only known avenue for outcrossing, which is estimated to occur in nature at a rate of about 1–2% (Cutter and Payseur 2003). Mechanistically, such males are generated periodically by non-disjunction events in the sex chromosomes (Cutter et al. 2003).

Gynodioecy (mixtures of hermaphrodites and females) is also present in extant invertebrate animals, albeit rarely, with cases known in the Porifera (sponges) and Cnidaria (anemones, corals, and allies) (Chornesky and Peters

FIGURE 3.7   The sea anenome *Epiactis prolifera*, a gynodioecious species.

1987; Jarne and Charlesworth 1993). Perhaps the first described case (Dunn 1975) involved a sea anenome, *Epiactis prolifera* (fig. 3.7). In this species, individuals begin life as females, but most specimens later become hermaphrodites if they survive long enough. Thus, this reproductive mode could also be termed protogynous simultaneous hermaphroditism.

Protandric Simultaneously Hermaphroditic Shrimps

For marine shrimp in the genus *Lysmata* (fig. 3.8), a typical individual begins its reproductive life as a functional male and later transforms into a functional simultaneous hermaphrodite (Bauer and Holt 1998; Bauer and Newman 2004; Bauer 2006). The gonads of these animals are ovotestes, the

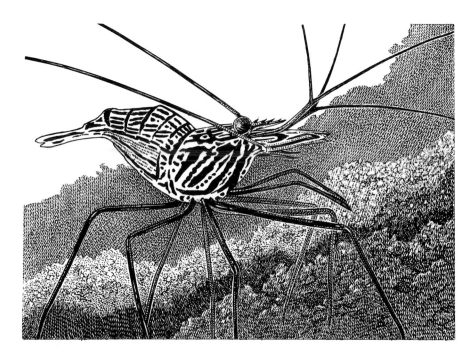

FIGURE 3.8   The Peppermint Shrimp, *Lysmata wurdemanni*, a protandric simultaneous hermaphrodite.

ovarian portion of which is slower to mature than the testis portion (Baeza et al. 2007). This reproductive lifestyle is termed protandric simultaneous hermaphroditism (PSH) or "adolescent protandry" (Ghiselin 1974), and it is also known in a polychaete worm (Premoli and Sella 1995) and land snail (Tomiyama 1996). Various hypotheses (not mutually exclusive) have been advanced to account for the possible adaptive significance of PSH: (a) a sex-dependent energetic cost model in which small, young individuals may have sufficient resources to promote the male function but insufficient re-sources to promote female function (de Jong and Klinkhamer 1994); (b) a sex-dependent mortality rate model in which young specimens that might reproduce as females or hermaphrodites would suffer higher mortality than young males (Klinkhamer et al. 1997); and (c) a time-commitment model, in which mortality rates are so high for small individuals that the longer time required to reproduce as a female precludes female function in the young (Day and Aarssen 1997). From laboratory manipulations of *Lysmata wurde-manni* in controlled experiments, Baeza (2006) concluded that models (a) and (c) were consistent with empirical observations for this species whereas model (b) was not.

Quasi-asexual Flatworms

Schmidtea mediterranea (fig. 3.9) is a planarian flatworm (phylum Platyhelminthes) that pushes the boundaries of bizarre reproduction. These simultaneous hermaphrodites copulate and exchange sperm, but the sperm cell usually does not contribute genetically to the offspring. Instead, it merely "pokes" an egg to trigger embryogenesis, such that most of the progeny arise asexually or clonally (D'Souza et al. 2004). This peculiar reproductive mode is known as sperm-dependent parthenogenesis or gynogenesis, which also occurs in several vertebrate taxa (Avise 2009). However, further genetic detective work has revealed that about 10% of the offspring in *S. polychroa* show evidence of sperm-mediated genetic exchange following copulations, such that the species really is quasi-sexual (or quasi-asexual) (D'Souza et al. 2004).

Other Oddities

Any list of hermaphroditic invertebrates with peculiar sexual morphologies and behaviors would be very long, but a few additional examples can give a flavor of the bizarre. In the land slug *Limax maximus*, hermaphrodites climb to an elevated perch, drop down on foot-long mucus strands, and exchange spermatophores (sperm packets) while dangling in aerial courtship (Langlois

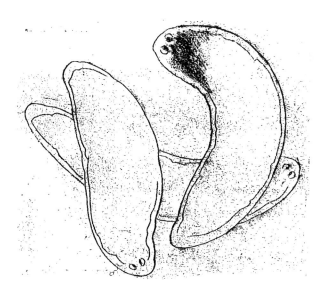

FIGURE 3.9   The planarian flatworm *Schmidtea mediterranea*, a hermaphrodite with a partially asexual (gynogenetic) reproductive mode.

1965). In various slugs in the genera *Ariolimax* and *Deroceras*, copulations sometimes are terminated by apophallation or amputation of the penis (Leonard et al. 2002; Leonard 2006), presumably as a energy-saving measure. In the marine flatworm *Pseudoceros bifurcus*, hermaphroditic individuals sometimes show "penis fencing," apparently to ward off damaging intromissions from partners (Michiels and Newman 1998). Finally, some hermaphroditic snails display "phally polymorphism" (Doums et al. 1998) in which a population can consist of individuals that are euphallic (have a fully developed penis), aphallic (lacking a penis), or hemiphallic (with a reduced penis) (Schrag and Read 1996; Viard et al. 1997). The penis condition influences an individual's mating behavior. For example, euphallic individuals may self-fertilize or outcross as either males or females (or both), whereas aphallic individuals can only self-fertilize or outcross as females.

Sex Determination and Pseudohermaphroditism

Collectively, invertebrate animals display diverse reproductive systems ranging from parthenogenesis (a unisexual form of clonal reproduction) to gonochorism (the presence of separate sexes) to sequential as well as simultaneous hermaphroditism (Coe 1943; Heller 1993). The mechanistic modes of sex determination are likewise highly diverse. Consider, for example, the mollusks, where various species displaying each of the following sex-determination mechanisms have been reported: a mammalian-like X/Y system in several gastropods (Vitturi et al. 1998) and the surfclam *Mulinia lateralis* (Guo and Allen 1994); an X/O chromosomal system in *Theodoxus* and *Littorina* snails (Vitturi and Catalano 1988; Vitturi et al. 1988, 1995); a *Drosophila*-like mechanism of X/autosome balance in the clam *Mya arenaria* (Allen et al. 1986); a suspected multi-locus genetic arrangement in the oyster *Crassostrea virginica* (Haley 1977, 1979); a system seeming to involve a dominant allele for maleness and another allele for protandric femaleness in the oyster *Crassostrea gigas* (Guo et al. 1998); and a system wherein maternal nuclear genotypes appear to influence gender in *Mytilus* mussels (Kenchington et al. 2002).

To add to this variety, environmental factors also sometimes impinge on sexual anatomies and functions in mollusks. For example, male whelks (genus *Busycon*) reared in the laboratory for many years have been reported to experience pronounced reductions in penis size (Castagna and Kraeuter 1994), and, analogously, trematode parasites can "castrate" mollusks in some cases (Køie 1969). High concentrations of the chemical tributyltin (used in anti-fouling paints) are known to impose anatomical sex changes on some gastropod mollusks (Gibbs et al. 1988; Power and Keegan 2001); for example, an "imposex" specimen may be a female initially, but, after exposure to the chemical, begin to display a vas deferens and a penis (although it is doubtful that these male sexual organs are fully functional; Power and Keegan 2001).

Jenner (1979) coined "pseudohermaphroditism" to refer to such environmentally induced switches between an apparent male phenotype and a female sexual phenotype in any invertebrate species that probably would be strictly gonochoristic under more normal circumstances. The discovery that particular environmental conditions sometimes precipitate changes of gender in mollusks or other invertebrates raises two important and interrelated distinctions: between hereditary (genetic) and environmental determinants of an individual's sex; and between genuine hermaphroditism and pseudo-hermaphroditism (Avise et al. 2004). In the remainder of this chapter, we will focus almost exclusively on individual invertebrate animals and species that are genuinely hermaphroditic in nature.

Evolutionary Histories: Gonochorism and Hermaphroditism

With respect to alternative expressions of dual sexuality, invertebrate animals tend to differ from plants (chapter 2) in at least two major regards. First, hermaphroditism is usually a derived (rather than ancestral) condition in invertebrate lineages, both overall and in many taxonomic subclades. This contrasts diametrically with the situation in many plant groups. Second, gynodioecy and androdioecy are both extremely rare in extant invertebrates. Thus, whereas gynodioecy in plants is quite common and seems to be a rather stable intermediary phase between cosexuality and dioecy, exactly how invertebrate animals transition between hermaphroditism and gonochorism is somewhat less clear. In many other regards, however, various expressions of dual sexuality in invertebrate animals bear close analogies to their counterpart phenomena in plants.

Phylogenetic Character Mapping

Hermaphroditism is undoubtedly polyphyletic in invertebrate animals, having arisen from ancestral gonochorism on many separate occasions. Eppley and Jesson (2008) used phylogenetic character mapping (PCM) to plot observed and deduced distributions of simultaneous hermaphroditism along external nodes and internal branches, respectively, of a phylogenetic tree for more than 80 invertebrate taxonomic classes (fig. 3.10). At this coarse evolutionary level of examination, hermaphroditism probably arose about a dozen times. This surely represents an underestimate, however, because closer inspection of particular clades indicates many additional origins for the phenomenon. For example, hermaphroditism probably arose on five different occasions in the nematode genus *Pristionchus* alone, in which five phylogenetically dispersed species display the phenomenon (fig. 3.11).

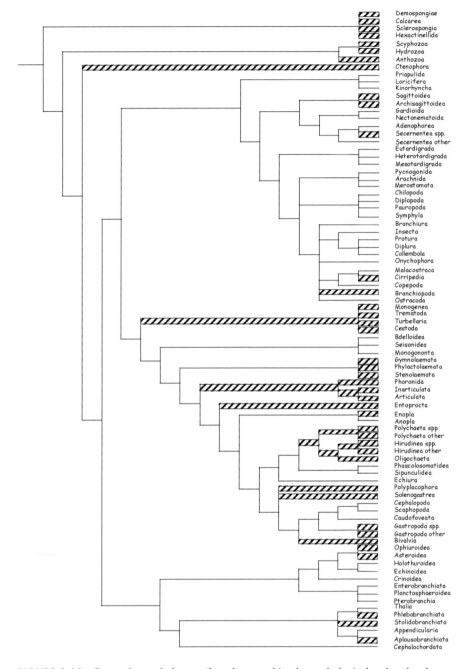

FIGURE 3.10 Coarse-focus phylogeny (based on combined morphological and molecular data) for major invertebrate taxonomic groups, suggesting multiple independent origins of simultaneous hermaphroditism (hatched bars) from gonochorism (after Eppley and Jesson 2008). The hermaphroditism phenomenon ranges from rare to ubiquitous across the various taxa and lineages highlighted by the hatched bars.

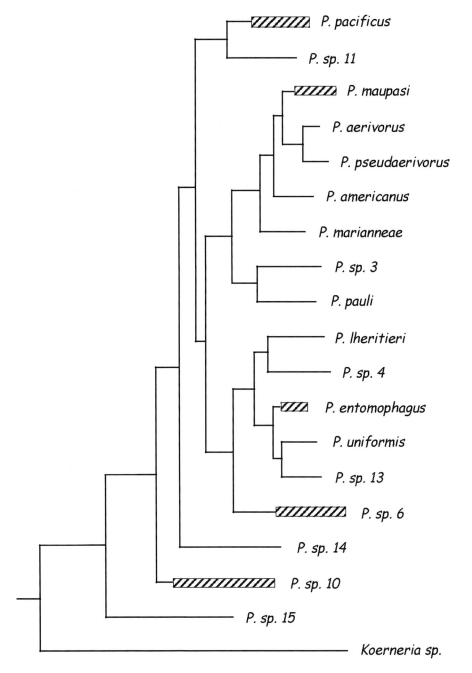

FIGURE 3.11 Phylogenetic tree (based on ribosomal genes) for various described and undescribed species of *Pristionchus* nematodes, indicating multiple origins of hermaphroditism (hatched bars) from gonochorism (after Mayer et al. 2007).

Thus, unlike the situation in flowering plants, the usual direction of the evolutionary shifts in the breeding systems of invertebrate animals appears to be from gonochorism (dioecy) to hermaphroditism. In another such example, a thorough phylogenetic assessment led Weeks and colleagues (2009) to conclude that dioecy was the ancestral breeding system in limnadiid clam shrimps and that there were at least two independent evolutionary transitions from gonochorism to hermaphroditism within the family. In general, such evolutionary shifts (plus, perhaps, a few changes in the reverse direction) seem to have occurred routinely in some invertebrate taxa (such as in some cnidarians, annelids, and mollusks; Ghiselin 1969; Clark 1978a), whereas strong phylogenetic constraints on breeding-system transitions seem to apply to other invertebrate taxa (Schärer 2002). Such evolutionary constraints might be due, for example, to a conservation of genetic or developmental processes that may underlie modes of sex determination in particular invertebrate clades (Uller et al. 2007).

Gynodioecy and Androdioecy

Both gynodioecy and androdioecy are rare (and probably ephemeral) conditions in invertebrates, thus making it difficult for PCM analyses to confirm that they are transitional evolutionary states between gonochorism and hermaphroditism. However, based on PCM, androdioecy itself probably arose from immediate ancestors that were hermaphroditic in most barnacles (Crisp 1983; Charnov 1987b; Høeg 1995), and from immediate ancestors that were gonochoristic in nematodes (Haag 2005). In clam shrimps, all-hermaphroditic populations or species may have had a direct dioecious ancestor in some cases (Weeks et al. 2006), but, on the other hand, they probably had an androdioecious progenitor on some other occasions (Weeks et al. 2009).

Pannell (2008, 2009) has emphasized a key potential difference between plants and invertebrate animals with respect to the functional consequences of androdioecy. In androdioecious plant species, hermaphrodites typically can outcross by mating with one another as well as with males, whereas in most androdioecious invertebrates (including rhabditid nematodes [Kiontke et al. 2004]) and clam shrimps [Sassaman and Weeks 1993]), hermaphrodites are physically unable to mate with one another, so males are the only viable source of outcross sperm. Thus, in contrast to the typical situation in androdioecious plants, androdioecious invertebrate species in which males are rare essentially may be confined to perpetual selfing, and this might lead (probably via inbreeding depression) to extensive lineage extinctions within their metapopulations (Pannell 2002; Loewe and Cutter 2008).

Genetics

For most invertebrate taxa, the precise genetic or biochemical mechanisms underlying the evolutionary transitions between gonochorism and hermaphroditism remain poorly known. However, nematodes in the genus *Caenorhabditis* (fig. 3.12) have long been model organisms for developmental research (Nigon 1949; Hodgkin and Brenner 1977), and their genetic and epigenetic bases of sex determination have been elucidated in great detail (reviewed in Haag 2005). *C. elegans* (like other nematodes) has an XX/XO sex-chromosome system in which the ratio of X chromosomes to autosomes is important in initiating a signal-transduction pathway that targets a key male-specifying gene (*mab-3*). At least 25 other male-promoting or female-promoting loci are also known to be involved in complicated metabolic pathways that help either to activate or repress spermatogenesis and oogenesis during germ-cell development, and thereby proximally mediate mechanistic transitions between male and female reproductive functions. Environmental factors such as nutrition can also impact sexual expression in *C. elegans* (Prahlad et al. 2003).

Self-fertile hermaphroditism has originated at least twice in different species of *Caenorhabditis* nematodes (Kiontke et al. 2004), and much has been

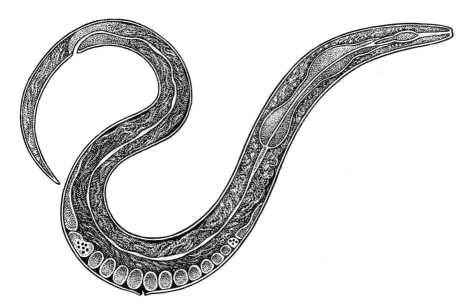

FIGURE 3.12   The nematode *Caenorhabditis elegans*, an androdioecious species that has been a model for genetic and developmental research.

learned about the particular mutations that probably effected the evolutionary transitions from gonochorism to dual sexuality (Nayak et al. 2005; Guo et al. 2009; Haag 2009). For example, the evolution of hermaphroditism in *C. elegans* appears to have occurred as a two-step process minimally involving: a mutation in the sex-determination pathway that initiated spermatogenesis in individuals that otherwise were genetically female; and another mutation that allowed the resulting spermatids to self-activate into viable sperm and thereby imbue the hermaphrodite with the capacity to self-fertilize (Baldi et al. 2009). In *C. briggsae*, the evolutionary transition to hermaphroditism was outwardly similar but the genetic underpinnings were rather different, involving mutational interventions at different points along the core sex-determination pathway (Hill et al. 2006). Thus, even within one small clade of nematodes, different suites of genetic changes apparently have precipitated convergent evolution of the observed hermaphroditic condition.

Similar sentiments will likely prove to apply to at least some other hermaphroditic invertebrates also, as the detailed genetic and physiological underpinnings of their sexual development become clarified through additional molecular detective work. Thus, Haag (2005:R978) appears to have been prescient when he concluded, "why and how crucial protein-protein interactions mediating sex determination are continually reinvented at the primary sequence level are interesting questions that will require the integration of structural biology, genetics, and ecology."

Selective Pressures

As described in chapter 2, a major challenge for botanists has been to understand why dioecy (gonochorism) evolved repeatedly from ancestral cosexuality in various plant lineages. For invertebrate biologists, an important challenge has been to understand why hermaphroditism evolved repeatedly from gonochorism in various animal lineages. This interesting reversal reflects the prevalence of gonochorism as an ancestral state in many invertebrates (both across higher taxa and in many younger clades), versus the prevalence of cosexuality as the probable ancestral state in many plant clades.

In the scientific literature for invertebrates, one long-standing hypothesis is that hermaphroditism tends to evolve when the resources that females can profitably allocate to ova are limited for some biological reason, such as a lack of brooding space, limitations in the time available for oviposition, or a restriction of oviposition to patchy microhabitats (Heath 1977, 1979). In such cases, female reproductive output may be constrained to a level below the maximum that available resources otherwise would allow, thus freeing "spare" resources that a female could redirect to produce sperm (ergo the evolution of hermaphroditism). This hypothesis has proved difficult to test critically, and some data seem inconsistent with it. For example, hermaphroditism is not disproportionately more common than gonochorism in coral species that brood their larvae compared to coral species with broadcast reproduction (Carlon 1999).

A hypothesis with perhaps greater support focuses on an organism's mate-finding efficiency. The idea is that when reproductive encounters or mate acquisitions are difficult (as in species that are sparsely distributed, rare, or sedentary), then hermaphroditism could be advantageous because of the fertilization insurance it provides. A selfing hermaphrodite need encounter no one to reproduce, and an outcrossing hermaphrodite need encounter only one other individual (as opposed to another specimen of the proper sex) (Tomlinson 1966; Jarne and Charlesworth 1993). Conversely, when mate acquisition is simple or energy-efficient, hermaphroditism should no longer be a stable evolutionary strategy. At least in theory, a mutation producing males or females should be able to invade a hermaphroditic population whenever selective pressures arise for the two sexes to specialize in different mate-search tactics (Puurtinen and Kaitala 2002).

Much circumstantial evidence can be interpreted as consistent with the fertilization advantages of hermaphroditism (Baker 1955; Pannell 1997). For example, in a geographic survey of the tadpole shrimp (*Triops cancriformis*), which is polymorphic for reproductive mode, Zierold and colleagues (2007) concluded that the postglacial colonization of northern Europe was facilitated by fertilization insurance, because northern populations are hermaphroditic or androdioecious whereas southern populations usually are gonochoristic. For tadpole shrimp (Notostraca) more broadly, hermaphroditism has been posited to facilitate long-distance dispersal, as these species have remarkable abilities to colonize isolated pools, even on oceanic islands (Longhurst 1955). In the limpet *Lottia gigantea*, which is a sequential hermaphrodite, population densities influence the probability of sex change in ways that appear to make sense in terms of enhancing fertilization success and individual fitness (Wright 1989). More generally, in a comparative study of multicellular organisms including many invertebrate species, Eppley and Jesson (2008) reported a statistically significant correlation (after corrections for phylogeny) between suspected mate-search efficiencies and breeding systems. All of these lines of evidence suggest that fertilization assurance is likely a key factor in the evolution of hermaphroditism from gonochorism in invertebrate animals.

Selfing Versus Outcrossing

Although the breeding systems of most hermaphroditic invertebrates remain poorly known (Bell 1982), sufficient evidence has accumulated to document that selfing is far from rare (Jarne and Charlesworth 1993; Jarne and Auld

2006). For many invertebrate species, reproductive mode can be surmised from an inspection of reproductive morphologies or behaviors. For example, a prevalence of ovarian tissue in the gonads of hermaphrodites often implies the occurrence of self-fertilization, because a high allocation to female reproduction is expected in selfing species (Charlesworth and Charlesworth 1981; Charnov 1982). For internal fertilizers, presence of an ovotestis (a gonad producing both eggs and sperm) often implies selfing, whereas hermaphroditic individuals that house male and female functions in separate organs are anatomically predisposed to outcrossing. With respect to behavior, hermaphroditic species in which individuals actively seek copulations (as in many terrestrial snails) are likely to be predominant outcrossers rather than selfers. However, exceptions to these tendencies do occur (see Doums et al. 1996 for clear examples involving freshwater snails), so final judgments about rates of selfing versus outcrossing in various species usually must await the application of molecular markers to deduce each population's breeding system via analyses of genetic parentage or population structure (see boxes 1.4 and 2.3).

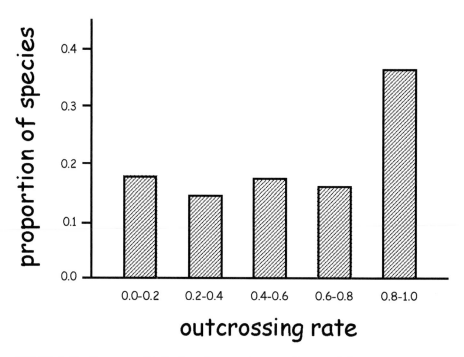

FIGURE 3.13 Frequency distribution of outcrossing rates (t) among 142 species of hermaphroditic invertebrate animals (after Jarne and Auld 2006). All estimates of t came from genetic markers as applied to progeny arrays or population structures in nature.

From a compilation of such data from the scientific literature, Jarne and Auld (2006) generated a frequency distribution of rates of outcrossing (t) and selfing ($s = 1 - t$) for nearly 150 invertebrate species. Many of the estimates proved to be intermediate between zero and one (fig. 3.13), a result that the authors interpreted as consistent with the hypothesis that selection pressures often favor the evolution and maintenance of mixed-mating systems. For particular invertebrate species, additional observations—such as the phenomenon of delayed selfing in relation to mate availability in freshwater snails (Tsitrone et al. 2003) and cestodes (Schjørring 2004)—also are consistent with the idea that selfing is a viable (albeit perhaps non-preferred) reproductive option for many simultaneous hermaphrodites. Another important point is that alternative mating systems probably are often facultative outcomes rather than genetically hardwired systems, in which case the issue shifts from how selective pressures might impact genetic polymorphisms for reproductive mode to questions about the fitness consequences of phenotypic plasticity for alternative reproductive patterns. In any case, mixed-mating systems are of special interest because they present fine opportunities to assess the fitness advantages and disadvantages of selfing.

Genetic Considerations

In theory, any allele that promotes selfing in an otherwise outcrossing population with no inbreeding depression has a 50% fitness advantage due simply to transmission genetics (Fisher 1941). Thus, selfing should be strongly favored unless counteracting selective forces also are at work. The major genetic deterrent to selfing is likely to be inbreeding depression (box 3.3). Given the transmission advantage that a selfing allele possesses, mean genetic fitness in the progeny of selfing must in theory be greater than 50% of the mean fitness of outcross progeny, if selfing is to evolve (Jarne and Charlesworth 1993). In other words, inbreeding depression must be < 0.5 for selfers to enjoy a fitness advantage over outcrossers. Theoretical models that assume a fixed level of inbreeding depression usually predict the evolutionary stability of either complete selfing or complete outcrossing, depending on the magnitude of the fitness reductions under inbreeding (Lloyd 1979; Charlesworth 1980). Such predictions do not mesh well with the common occurrence of mixed-mating systems in hermaphroditic invertebrate species.

The situation in real life is much more complicated, however, in part because the magnitude of inbreeding depression itself can change through time. Inbreeding depression tends to decrease under selfing when deleterious recessive alleles are exposed routinely to natural selection and thereby purged from populations (as should often be true in species with high rates of selfing). At least in theory, populations with intermediate or high incidences of selfing can be evolutionarily stable when the rate of inbreeding

BOX 3.3 Inbreeding Depression in Invertebrate Animals

The empirical case for widespread inbreeding depression in invertebrate animals appears to be less well developed than for plants and for vertebrate animals (Jarne and Charlesworth 1993; Knowlton and Jackson 1993). For example, a major review of inbreeding depression in the wild (Crnokrak and Roff 1999) included 157 data sets for 15 plant species and 18 vertebrates but only one invertebrate species (the land snail *Arianta arbustorum*; Chen 1993). Some additional reports of inbreeding depression in invertebrates involve a marine ascidian *Corella willmeriana* (Cohen 1996), the scallop *Pecten maximus* (Beaumont and Budd 1983), the housefly *Musca domestica* (Bryant and Reed 1999), the fruitflies *Drosophila melanogaster* and *D. virilis* (Frankham, 1995), freshwater snails in the genus *Physa* (Wethington and Dillon, 1997; Jarne et al. 2000), and nematodes in the genus *Caenorhabditis* (Dolgin et al. 2007).

The relative paucity of direct evidence for inbreeding depression in invertebrates is probably due to a general shortage of suitable scientific studies rather than to rarity of the phenomenon. Consistent with this notion is the fact that many species of simultaneously hermaphroditic invertebrates have evolved genetic self-incompatibility systems, analogous to those in many plants, that inhibit self-fertilization via sperm-egg interactions (Scofield et al. 1982; Carlon 1999; Bishop and Pemberton 2006). On the other hand, many invertebrate species with simultaneous hermaphroditism have mixed-mating systems with a strong self-fertilization component, and at least some of these display little or no obvious inbreeding depression (e.g., Doums et al. 1996). Thus, any fitness difficulties that might be associated with inbreeding in hermaphrodites are neither universal nor insuperable. One likely explanation involves an amelioration of inbreeding depression via the purging of deleterious alleles in populations that routinely self-fertilize.

Measuring inbreeding depression in natural populations is a difficult task for many reasons (Jarne and Charlesworth 1993), not least being the desirability of integrating components of fitness at all phases of the life cycle, including estimates of both fertility and viability. Few such thorough analyses exist for invertebrate animals, but one fine example involved the freshwater snail *Physa acuta* (Jarne et al. 2000). In this hermaphroditic species, inbreeding depression from selfing was severe (> 0.9), as might be expected for this highly outcrossing species ($t = 0.9$).

and a relief from inbreeding depression coevolve (Charlesworth and Charlesworth 1990; Barrett and Charlesworth 1991; Uyenoyama et al. 1993; Lande et al. 1994). Inbreeding depression also can increase across time, under selfing, when fitness is due to overdominant loci, i.e., those with heterozygous advantage (Charlesworth and Charlesworth 1987b; Ziehe and Roberds 1989). In some such cases, intermediate selfing rates again can be evolutionarily stable (Charlesworth and Charlesworth 1990). Finally, mixed-mating systems may be rather immune from loss when inbreeding depression is close to 0.50, because alleles for selfing or outcrossing then tend to act as if selectively neutral and are not easily driven to loss or fixation by natural selection (Charlesworth et al. 1990, 1992).

Apart from such genetic considerations, many ecological factors also come into play in determining the fitness consequences of selfing versus outcrossing. For invertebrate animals (as is also true for plants; chapter 2), the most widely discussed of these factors are related to the concept of fertilization insurance.

Ecological Considerations

Assurance of fertilization is greater for self-compatible hermaphrodites than for hermaphrodites that obligately outcross. Conventional wisdom is that outcrossing normally is favored in simultaneous hermaphrodites for reasons related to inbreeding depression, but that selfing can evolve under special ecological circumstances when choices are necessitated between self-fertilization and perhaps no reproduction at all (Lloyd 1979; Holsinger 1996). For example, selfing could be highly advantageous for invertebrates that routinely experience difficulties in finding mates. Such may often be true in species that are rare, sedentary, patchily distributed, or prone to colonize new habitats.

Various lines of evidence for invertebrate animals are consistent with the notion that fertilization insurance is a key factor impacting the evolution of selfing rates in simultaneous hermaphrodites. Across gastropod mollusks, selfing seems to be correlated with colonization propensity (Selander and Ochman 1983), which makes sense if colonizers routinely find themselves alone in newly opened habitats. Such findings give further circumstantial support for Baker's rule (chapter 2), which states that the capacities for self-fertilization and long-distance dispersal are causally related across species because of the reproductive assurance that automatically comes with being a selfing hermaphrodite. Selfing is also common in pulmonate gastropods with patchy distributions and limited natural dispersal capabilities (e.g., Foltz et al. 1982, 1984; Jarne et al. 1993), presumably because these factors also contribute to limited mating opportunities. (An alternative explanation, however, is that limited dispersal leads to biparental inbreeding—via matings between related individuals—which in turn leads to a purging of genetic load, which in turn opens wider windows of opportunity for successful selfing.) More generally, simultaneous hermaphroditism in invertebrates traditionally has been associated with low population densities, low vagilities, and parasitic lifestyles (reviewed in Schärer, 2009).

Many additional ecological considerations could also impact the genetic fitness of selfers versus outcrossers. For example, outcrossing could be detrimental to individuals in species with high incidences of sexually transmitted diseases (STDs); and even when STDs are absent, outcrossers have the added costs of time and energy required to find and woo suitable mates (Puurtinen and Kaitala 2002; Eppley and Jesson 2008).

Joint Genetic and Ecological Considerations

Continued selfing dramatically reduces genetic variation (heterozygosity) within a lineage, and thereby has an additional effect of reducing effective genetic recombination (see box 2.6). This in turn allows, in principle, the buildup of homozygous coadapted gene complexes (potentially throughout the genome) that otherwise would be broken apart in each generation of outcrossing. The consequences of this phenomenon were described in chapter 2 for the evolution of multi-locus genomic adaptations to local ecological conditions in the slender wild oat, *Avena barbata*, which has a mixed-mating system with predominant selfing and occasional outcrossing.

The hermaphroditic land snail *Rumina decollata* (fig. 3.14) similarly displays a mixed-mating system with predominant selfing, and this invertebrate provides a classic demonstration of the potential for adaptive convergence in the genetic architectures of selfing animals and plants (Selander et al. 1974; Selander and Kaufman 1975). This snail is native to the Mediterranean region where its populations exist as a complex of highly inbred or mildly outcrossed strains characterized by different suites of allozyme alleles and morphological markers. In surveyed areas of southern France, two strains predominate: a dark form that occupies protected mesic environments under logs or rocks, and a light form associated with open xeric habitat. These two types also showed fixed differences at 13 of 26 allozyme loci. Further analyses revealed that occasional outcrossing between these strains releases extensive recombinational variation that otherwise is expressed as between-strain genetic differences. Nonetheless, these two forms of snail tend to retain their separate identities and habitat correlations in nature.

These findings suggest that, as in *Avena barbata*, strong multi-locus associations and pronounced population genetic structures in this snail species probably stem from natural selection (as well as stochastic population factors; Selander 1975) that operates in conjunction with self-fertilization to reduce effective recombination (Selander and Hudson 1976). In such cases,

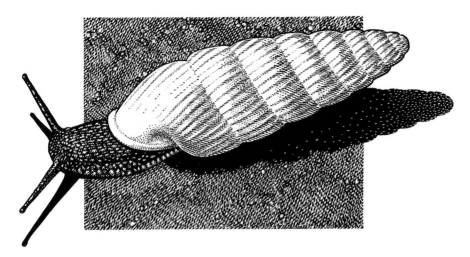

FIGURE 3.14   The hermaphroditic land snail, *Rumina decollata*.

selfing is not merely a reproductive option of last resort, but rather an integral and adaptively significant component of the mixed-mating system.

As an additional note of interest, *R. decollata* was introduced to the eastern United States before 1822 and subsequently expanded across much of the continent. All assayed populations in North America proved to be identical in allozyme composition and, thus, appear to be derived from a single strain introduced from Europe (Selander and Kaufman 1973). These findings indicate that despite an absence of appreciable genetic variation, some highly inbred strains of *R. decollata* can enjoy great ecological success (wide distributions and high population numbers), at least in the evolutionary short term. They also illustrate the idea that fertilization insurance and the potential for maintaining coadapted multi-locus genotypes are not mutually exclusive explanations for the adaptive significance of selfing in hermaphroditic species that are adept at colonization. Instead, these two routes to fitness enhancement can be synergistic.

Sex Allocation in Simultaneous Hermaphrodites

For mathematical tractability, formal sex allocation theory (SAT) for hermaphrodites traditionally made several simplifying assumptions: (a) all individuals in a population have the same fixed or zero-sum budget for lifetime reproduction, with a linear trade-off occurring between resources invested in maleness versus femaleness; (b) this budget is separate from

other life-history traits; (c) shifts of investment into alternative sex functions can be related by simple equations to fitness gains, as depicted in
"fitness-gain curves" (box 3.4); and (d) generations are nonoverlapping. SAT
makes numerous predictions that have been subject to observational and
experimental tests involving invertebrate animals. Indeed, probably more
scientific literature has been devoted to issues of sex allocation than to any
other conceptual topic related to invertebrate mating systems (Charnov
1982; Schärer 2009). Examples of SAT predictions and their empirical eval-
uations will be described next. Some of these hypotheses conceptually
overlap.

BOX 3.4   Fitness-gain Curves

Fitness-gain curves relate a simultaneous hermaphrodite's reproductive
fitness to its investments into male versus female function, and thus they
lay much of the groundwork for sex allocation theory (SAT). Each curve is
a power function (described by an equation of the form $y = ax^b$) in which
the exponent b determines the curves shape: $b = 1$, linear relationship;
$b < 1$, diminishing curve; $b > 1$, accelerating curve (panel A).

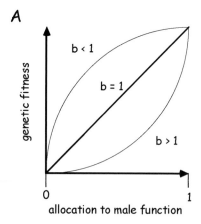

Fitness-gain curves can differ in shape for male versus female functions, and their shapes predict how reproductive resources optimally
should be allocated between these two roles. In panel B, for example, the
curve for male fitness is convex (eventually shows diminishing returns
with increasing investment in male function), whereas the line for female
fitness is straight (increases linearly with increasing investment in female
function). An individual's overall fitness is optimized at the curve's inflec-

(continued)

BOX 3.4 (*continued*)

tion point (*m*) between expanding and diminishing returns on male investment, with female investment then being $f = 1 - m$. Panel C shows an alternative representation of this situation, but in this case emphasizing that the sum of the male and female fitness curves identifies the individual's optimal sex allocation. In this panel, the x-axis is to be interpreted as the percentage of an individual's resources devoted to male function, with the remainder being devoted to female function.

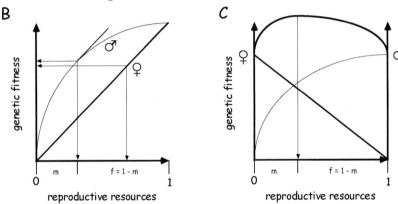

The shapes of these curves for particular species or biological settings depend on key assumptions about how different investment tactics might translate into fitness gains. For example, panel D shows male fitness-gain curves at they might relate to variation in mating group size under the local mate competition (LMC) model and panel E shows female fitness-gain curves as they might relate to brooding constraints (see text). Panel A is redrawn from figure 1 in Baeza (2007), and panels B through E are based on figure 1 in Schärer (2009).

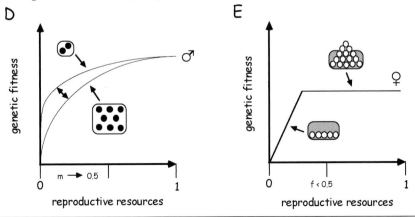

Selfing and Allocation to Female Function

SAT generally predicts that the evolution of increased selfing rates should be accompanied by increased relative allocations to female reproduction function (Charlesworth and Charlesworth 1981). This prediction follows from the notion that as the mean selfing rate in a population increases, an individual's opportunity to improve its fitness via "sperm broadcasting" is reduced; and the energy saved by reducing investments in sperm presumably can be spent on fitness-enhancing female functions such as egg production or care of young. Furthermore, because of anisogamy, even a relatively small investment in sperm should suffice to fertilize all of the available eggs within a focal selfer.

Johnston and colleagues (1998) provided the first empirical test of this SAT prediction in an invertebrate species. For the freshwater mussel *Utterbackia imbecillis*, the authors found, as predicted by SAT, that individuals in populations with higher selfing rates did indeed devote significantly more reproductive tissue to female functions and less to sperm production. This finding parallels observations in plants, where relative allocations to male function tend to decline as the selfing rates increase (Cruden 1977; Lovett-Doust and Lovett-Doust 1988).

Stress and Allocation to Male Function

Especially for species that occupy patchy habitats that vary greatly in quality, SAT generally predicts an increased allocation to male function for individuals confined to the more stressful settings (Freeman et al. 1980a,b). This follows from the idea that even minimal female function is energetically expensive compared to male function, so that when faced with special stresses, an individual tends to fare best by reproducing mostly as a male. Such stresses could be of many sorts, ranging from physical conditions such as temperature to biotic challenges such as parasites (Calado et al. 2005).

Hughes and colleagues (2003) tested this prediction in a sessile marine hermaphrodite, the colonial bryozoan *Celleporella hyalina*. Consistent with SAT, the authors found a stress-induced bias toward male function in colonies experiencing impaired prospects for parental productivity or survival. The response was phenotypically plastic (as shown by exposing cloned replicates of particular colonies to different environments), and it parallels the kinds of facultative shifts toward maleness that have been reported widely in cosexual plants exposed to diverse environmental stresses (Freeman et al. 1980a,b).

Local Mate Competition

Competition between male gametes is often a key factor in an individual's reproductive success. This realization laid a foundation for the "local mate competition" (LMC) model of Hamilton (1967), which predicts a relationship in gonochorists between the size of the mating group and relative investment in male reproductive function. It was Hamilton's insight to see that when brothers compete for mates, they become, in effect, genetically redundant with regard to transmitting their mother's genes, and in such cases LMC favors mothers who can adjust their sex ratio toward daughters. In other words, as fewer and fewer mothers contribute eggs to a local mating group, the intensity of local mate competition rises, and selection should reward sex-ratio shifts toward daughters because such shifts tend to reduce futile conflict among related sons (Taylor 1981).

A similar situation arises in hermaphrodites when competition for fertilization occurs between related sperm cells (Greeff et al. 2001; Schärer and Wedekind 2001). Consider, for example, the extreme case of a monogamous simultaneous hermaphrodite, whose sperm have a mean coefficient of genetic relatedness of $r = 0.5$ (the same level of r as that of brothers in a gonochoristic species). From the sperm donor's perspective, the optimal strategy would be to produce only enough sperm as required to fertilize all of the partner's eggs, and to invest otherwise in egg production (leading to a female-biased SA). As the monogamy assumption is relaxed and matings become more promiscuous, mean r in the pool of competing sperm decreases, and it behooves each male to invest relatively more in sperm production, up to a maximum of 50%. Thus, in general as applied to hermaphrodites, LMC predicts that as the mating-group size increases, an individual's allocation to male function should also increase (Charnov 1982; Lloyd 1984; Lively 1990). Schärer (2009) suggests that this phenomenon should be termed local sperm competition (LSC) to distinguish it from LMC in gonochorists.

Trouvé and colleagues (1999) found support for predictions of the LMC (or LSC) model in their studies of a parasitic trematode (phylum Platyhelminthes), *Echinostoma ilocanum* (fig. 3.15), a simultaneous hermaphrodite that invades and mates in the intestine of vertebrates. In analyses of mice experimentally infected with one, two, or 20 metacercariae (larvae), the authors found that the parasite's relative allocation to male reproductive function adjusted to lower levels in the smaller-sized mating groups. Several additional studies of hermaphroditic invertebrates likewise have reported phenotypic adjustments of SA in response to variation in size of the mating group: in barnacles (Raimondi and Martin 1991; Hoch 2008, 2009), and in some flatworms and leeches (Schärer and Ladurner 2003; Tan et al. 2004; Brauer et al. 2007).

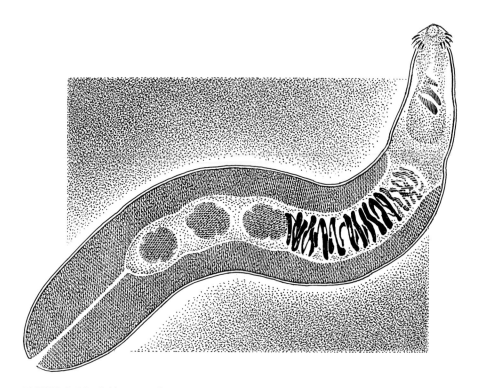

FIGURE 3.15   *Echinostoma ilocanum*, a hermaphroditic parasitic nematode.

Because most of these responses were facultative (i.e., phenotypically plastic), they also provide a useful reminder that many of the predictions of SAT apply not only to genetic changes in populations across evolutionary time but also to physiological adjustments of individuals to current ecological conditions. Anthes and colleagues (2006a,b) argue that selection often should favor flexible adjustments of mating role preferences also, as opposed to fixed population-specific preferences as sometimes is assumed (Leonard and Lukowiak 1984; Leonard 1999). Indeed, phenotypic plasticity for both sex allocation and mating options may be among the primary selective advantages of hermaphroditism itself (Michiels 1998).

Local Resource Competition

The LRC model of sex allocation assumes that related offspring (such as half sibs) often compete for limiting resources in the local environment, and thereby restrain the reproductive success of their parents (Clark 1978b; Charnov 1982). If so, then such competition can produce selection pressures that might favor heavier parental investment in the sexual function with higher gene dispersal. Male and female gametes are often differentially dispersive, as for example in wind-pollinated plants and in various marine invertebrates that broadcast sperm but brood eggs and larvae. If larger individuals produce more seeds or eggs, and if these are far more sedentary than male gametes, then LRC could promote a shift toward increased male allocation with increasing size of the individual, all else being equal.

Schärer and colleagues (2001) addressed the LRC scenario as a possible contributor to observed patterns of sex allocation in the tapeworm *Schistocephalus solidus*. They concluded in this case, however, that LRC (in contrast to LMC) was not likely to be a major selective force favoring size-dependent sex allocation, because the fertilization regimen (inside the host) effectively forces similar dispersal patterns for male and female gametes.

Brooding Constraints

For simultaneous hermaphrodites that brood their eggs, the available brood space in an individual presumably sets an upper limit on optimal allocation to female function (see box 3.4). Above that boundary, no further gains in fitness are expected with further investments in female operations. Thus, a male-biased SA is predicted whenever the brood space can be filled with less than 50% of an individual's reproductive resources (Charnov 1982). Empirically, however, few studies have estimated SA in brooding species, and much of the limited available evidence seems at face value either equivocal (as in corals; Hall and Hughes 1996) or contrary (as in starfish [Strathmann et al. 1984] and shrimp [Baeza 2007]) to any predicted tendency for male-biased SA in brooding invertebrates.

Intra-individual Trade-offs

For hermaphroditic individuals, traditional SAT assumes that the allocation of reproductive resources to one sexual function inevitably reduces the amount of resources available for the alternative sexual role. Optimal sex allocation then depends on how the relative energetic investments in alternative sexual functions translate into returns on investment (reproductive success). Allocation trade-offs seem physiologically plausible if not inevitable, but they have proved difficult to confirm empirically (Schärer 2009). Although some studies have proffered data in support of SA trade-offs in invertebrate animals, including pond snails (DeVisser et al. 1994; but see also Koene et al. 2006), colonial ascidians (Yund et al. 1997), and polychaete worms (Sella and Lorenzi 2003), other studies have found no correlation (marine shrimp, Baeza 2007; land slugs, Jordaens et al. 2006), equivocal evidence (flatworms, Schärer et al. 2005), or even a counterintuitive positive correlation (land snails, Locher and Baur 2000) between levels of allocation to male and female functions. Of course, weakness of evidence does not necessarily evidence weakness for the assumption of allocation trade-offs. Thus, the current paucity of documentation for trade-offs might just be attributed to the many methodological challenges of empirically quantifying sex allocation in hermaphrodites (box 3.5).

BOX 3.5   Male Versus Female Allocation Trade-offs: Real or Not?

Two fundamental assumptions of traditional sex allocation theory (SAT) are that all hermaphroditic individuals in a population have a reproductive budget that is separate from the energy budgets for other traits, and that the reproductive budget entails a linear trade-off (a zero-sum game) between resources invested in male versus female functions (see the figure below, modified from Schärer 2009). These critical assumptions, although perhaps intuitively appealing, have proved difficult to confirm with empirical data.

reproductive resources		resources for other life history traits		
male (m) ←→	female (1 - m)	trait 1	trait 2	trait 3

Much of the difficulty may lie in measuring sex allocation reliably, a challenge that has several facets as detailed by Schärer (2009):

(a) *variable resource costs.* Variable fertility within and among individuals has been the focus of most SA studies, with the costs (resources allocated to male and female gamete production) typically measured by surrogate criteria such as gonadal dry weight or volume. A serious but often overlooked possibility is that the cost of producing or maintaining a unit of testicular tissue, for example, might be different from that for a unit of ovarian tissue.

(b) *fixed reproductive costs.* Primary sexual characters (notably the genitalia) are often considered fixed costs that all reproductive animals must pay, regardless of fertility. However, genital morphology itself apparently can evolve by sexual selection and sexual conflict (Hosken and Stockley 2004), thus indicating genetic variation for such traits. Such between-individual variation in the costs of building and maintaining primary sexual traits probably should be (but typically has not been) included in empirical appraisals of SA.

(continued)

BOX 3.5   (*continued*)

(c) *integration across time.* Most empirical studies of SA have entailed only point estimates of reproductive costs, and these may often be misrepresentative of an individual's investments as integrated across longer periods of time (Schärer et al. 2004).

(d) *measuring function as both sexes.* SA sometimes is measured for one sexual function only (as for example in many species for which it is easier to count ova than sperm), and then to assume that any change in allocation to that sex registers an SA response. This clearly is inappropriate, because it merely presumes a pattern that should be critically tested.

(e) *absolute versus relative measures.* For all of these and other reasons, quantifying an individual's absolute investments of energy and resources to male versus female functions can be daunting. Fortunately, questions about the relative investments in male versus female function (for example, is the SA of individual A more female-biased than that of individual B?) tend to be more tractable yet still can be highly informative in many circumstances.

A related issue is whether sexual adjustments *per se* might be functionally costly to simultaneous hermaphrodites. If so, individuals often might display suboptimal allocations to male versus female functions if the cost of adjusting sexual function exceeds the cost of maintaining the otherwise suboptimal tactic. Few experimental studies have addressed this issue, but Lorenzi and colleagues (2008) report that sexual adjustments are not costly in the polychaete worm, *Ophryotrocha diadema*.

Sexual Conflicts

For outcrossing simultaneous hermaphrodites, the mating partners may not be in lock-step with regard to optimal reproductive tactic. SAT and game theory predict that conflicts over optimal SA routinely arise between sperm donor and sperm recipient. Specifically, whereas a recipient's fitness ideally should become adjusted to maximize its total fitness as sire and dam, a donor's fitness would be maximized if it could coax or coerce the recipient to allocate most of its resources to female function (Charnov 1979; Michiels 1998). Thus, SAT predicts that sperm donors should be under selection for traits that enhance female function in the recipient, and recipients should be under selection for traits that counteract attempted manipulations by the sperm donor. Possible examples of such sexual conflict already have been mentioned: the love-dart apparatuses of snails and the skin-piercing setae of earthworms. Furthermore, in pond snails and land snails, evidence indicates that the allohormones delivered by the darts affect not only sperm uptake or utilization but also can alter patterns of sex allocation in the recipient (Koene 2006).

Opposing selective pressures for the hermaphroditic participants in outcross matings provide an example of what is known as sexually antagonistic coevolution, an important category of sexual conflict (Chapman et al. 2003; Arnqvist and Rowe 2005; Parker 2006). There are many hypothetical routes toward resolving such sexual conflicts. One example—gamete trading—is described in box 3.6.

BOX 3.6   Gamete Trading

In gonochoristic species and in sequential hermaphrodites, the decision on who donates sperm and who receives sperm is dictated by the current sex of the mating partners. In simultaneous hermaphrodites, by contrast, each mate in principle could insist on mating either as male, female, or both, with the choice potentially having a large impact on genetic fitness. This can lead to conflicts of interest over gender roles, and to behavioral resolutions of such conflicts. "Sperm trading" may illustrate one such compromise. The basic idea, first formulated by Leonard and Lukowiak (1984) for an internally fertilizing species (the sea slug *Navanax inermis*), is that each simultaneous hermaphrodite conditions sperm donation upon sperm receipt.

Observations deemed supportive of the sperm-trading hypothesis were made on the sea slug, *Chelidonura hirundinina* (Anthes et al. 2005). In this species, a single mating sequence involves several successive penis insertions and inseminations per partner, with the first copulation typically involving simultaneous sperm donation and subsequent copulations involving alternating unilateral inseminations. The researchers were able to manipulate the sea slugs such that sperm transfer in the initial mating was blocked but copulation behavior was otherwise unaffected. In this experimental regime, individuals proved to be reluctant to inseminate a partner who was nonreciprocating with respect to sperm donation. However, in another sea slug species (*Chelidonura sandrana*), similar experiments failed to uncover strong evidence for sperm trading as a means of enforcing reciprocal insemination (Anthes and Michiels 2005).

A somewhat analogous process—egg trading—might apply as well to simultaneous hermaphrodites that shed their eggs and have external fertilization (Fisher 1980; Sella 1985; Sella et al. 1997). However, sperm trading and egg trading have a key potential difference: egg trading normally proceeds to reciprocal fertilizations, whereas sperm traders with internal

(continued)

BOX 3.6   (*continued*)

fertilization may have limited control over the fate of their sperm in their partner. This raises a potential conundrum, and suggests that sperm trading sometimes might be impacted by selective factors in addition to fertilizations *per se*, such as a recipient's digestion of sperm cells for nutrition (Pongratz and Michiels 2003; Anthes and Michiels 2005).

Although much has been written about gamete trading (e.g., Vreys and Michiels 1998; Michiels and Bakovsky 2000; Landolfa 2002; Koene and Ter Maat 2005), the topic remains rather controversial, both in concept and with respect to critical empirical support. One of the difficulties is that the straightforward observation of reciprocal gametic exchange by itself is insufficient to prove that the transfers were conditional as opposed to unconditional.

Anthes and colleagues (2006b) survey the wide variety of conceptual approaches (often nuanced and overlapping) that have been developed to predict the outcomes of gender conflicts between outcrossing hermaphrodites: sex allocation models, risk-aversion models, opportunistic-male models, and game-theoretical models. SA models tend to focus on physiological gender adjustments that may take many hours or days in most hermaphroditic invertebrates (e.g., Schärer and Ladurner 2003); but these models may be less germane when an individual can make more-or-less instantaneous mating decisions about how to behave vis-à-vis a specific partner. Risk-aversion models assume that mating systems are species-specific and that the optimal tactic for an individual is to maximize control over the fertilization process (i.e., to minimize the fertilization risk) (Leonard and Lukowiak 1984; Leonard 2005); but these assumptions may not always hold. The opportunistic-male models focus on how hermaphrodites should become increasingly choosy, as mating rates and sperm competition increase, about whom should receive their sperm (Greeff and Michiels 1999; Michiels et al. 2003); but these models may be less apropos when mating rates are low, as they are in many hermaphroditic species (Ghiselin 1969). Finally, the game-theoretical models typically assume that mating partners can either cooperate by reciprocally performing both sex functions, or cheat by adopting only the preferred sex role (Fisher 1980; Axelrod and Hamilton 1981; Sella 1988); but again the assumptions can be challenged in some cases, such as when an individual can choose from a continuum of tactics intermediate between full cooperation and untempered cheating (Killingback and Doebeli 2002; Doebeli and Hauert 2005).

In their review of the relevant literature, Anthes and colleagues (2006b) conclude that although each of these classes of models can yield some insights, none by itself satisfactorily explains the rich diversity of mating systems observed across hermaphroditic species. The authors propose that most hermaphroditic mating systems are extremely plastic, capable of continuously shifting back and forth between, for example, male-biased and female-biased sex role preferences, unilateral and reciprocal copulations, conditional and unconditional reciprocity, or standard copulations versus hypodermic impregnations. To the extent that such sentiments are valid, simultaneous hermaphrodites have even more fitness-enhancing options and far greater reproductive flexibility than do most gonochoristic species, and these benefits apparently outweigh whatever costs are associated with the continuous maintenance of two reproductive systems per individual.

Other Factors

In general, SAT predicts that hermaphrodites optimally divide their resources between male and female functions, but the particular tactics actually adopted in a given species are likely to vary widely as a function of many individual-level and population-level selective pressures (not to mention phylogenetic history). To pick just one example of the potential complexity of factors impinging on reproductive tactics, Angeloni and colleagues (2002) developed a theoretical model that incorporated several traits of an individual, its current mate, and the population. Their model showed that optimal SA can depend on such factors as the body-size distribution in the population, the size of an individual relative to its mate, the disparity of resources between size classes, the cost of filling a sperm storage organ, and the shape of the sperm displacement function, all of which can affect the shape of the fitness-gain curves (e.g., Yund 1998). The net theoretical ramification is that plausible variation in numerous biological parameters can lead to widely varied expectations about optimal sex-allocation tactics in a population.

Sex Allocation in Sequential Hermaphrodites

With respect to sex allocation, sequential hermaphrodites can be thought of as simultaneous hermaphrodites that have gone to the extreme lengths in their ontogenetic shifts between male and female reproductive function. Depending on the species, some individuals switch entirely from male to female, some from female to male, some switch back and forth, and some may decline to switch at all (Policansky 1982). Thus, SAT again is relevant, in this case endeavoring to explain when and why individuals fully change gender (Munday et al. 2006). The theory can apply to how natural selection might shape evolutionary-genetic outcomes in populations, or to how phenotypic flexibility in sexual development might pay fitness dividends to individuals that can respond appropriately to relevant factors in their immediate environments.

For some invertebrate species including particular shrimps (Bergström 1997) and snails (Branch and Odendaal 2003), age or body size at the time of sex change seem to be genetically programmed because these switches occur almost regardless of environmental circumstance. But in other species including some polychaete worms (Berglund 1986), shrimps (Charnov and Hannah 2002; Baeza and Bauer 2004) and snails (Collin 1995; Warner et al. 1996; Chen et al. 2004), the timing of sex change is known to be a flexible response by individuals to local conditions. For various sex-changing invertebrate species, such local conditions may include the size of an individual relative to others in the social group (Warner et al. 1996), local population density (Wright 1989), and the sex ratio in the immediate social assemblage (Collin et al. 2005).

The Size-advantage Hypothesis

When considering sequential hermaphrodites from the perspective of sex allocation theory, the size-advantage hypothesis (SAH) has been paramount (Ghiselin 1969; Warner 1975, 1988). This hypothesis will be developed at greater length in chapter 4 (because it has been applied extensively to fishes), but the basic idea is as follows. In most species, the sizes and ages of individuals are positively correlated (because specimens tend to grow during their lifetimes). The SAH predicts that sex change is favored by natural selection when an individual reproduces most effectively as one sex when small and young, but as the other sex when larger and older. Depending on the biology and ecology of a particular species, males might have the reproductive advantage when small and females when large, in which case protandry (male-first hermaphroditism) could often be favored by natural selection. In many other species, individuals might maximize their fitness by functioning as females when small or young but as males when larger and older, in which case protogyny (female-first hermaphroditism) should be favored. The empirical challenge is to understand what biological conditions might tip the scales in general favor of individuals reproducing as dams versus sires at various size cohorts or age classes.

One important consideration is the energetic cost of reproduction. When even minimal function as a female requires a threshold of energy investiture that lies beyond what a small or young individual can reasonably expend, youthful maleness may be favored by natural selection. Many protandrous invertebrate species may illustrate this phenomenon. In such species, small and young individuals simply may be incapable of reproducing as a female, yet they have the capacity to sire at least some progeny by producing and deploying sperm. Later, as such individuals grow, they may finally cross a threshold that permits the addition of female functions to their reproductive repertoire.

In a comparative survey of hermaphroditic reef-building corals, Hall and Hughes (1996) found striking differences in sex allocation that appeared to relate to the energetics of fertilization biology in these modular colonial organisms. In particular, they found that young colonies invest disproportionately in testes, thus allowing reproduction to commence without a heavy initial expense of egg production, while also buying time for the colony to grow to a larger and safer size.

Another reason why maleness is often disproportionately advantageous early in life relates to the correlation between body size and fecundity. Whereas even small males often can produce prodigious quantities of sperm, in many species only large individuals can produce big clutches of eggs. Thus, all else being equal, natural selection often should bias toward male allocation early in an animal's life and more toward female allocation as the animal grows. Schärer et al. (2001) tested this prediction in the parasitic tapeworm *Schistocephalus solidus*, a simultaneous hermaphrodite in which adult size is highly variable. They found that patterns of sex allocation were related strongly to body volume, with larger individuals significantly biased toward female function.

In some other invertebrates, youthful reproduction as a female may be favored by selection. SAT generally predicts this pattern for species in which older, larger, or more experienced males have a greater wherewithal to attract mates or monopolize mating events, and thereby achieve disproportionate reproductive success. Such appears to be the case in some freshwater and intertidal isopods (fig. 3.16). In these protogynous hermaphrodites, males actively guard their mates and thereby prevent stolen copulations (cuckoldry) by other males. Larger males guard females with larger broods and thereby tend to realize higher genetic fitness. Several studies on isopods have concluded that mate guarding is a primary selective pressure favoring protogyny in these invertebrate animals (Brook et al. 1994; Abe and Fukuhara 1996).

Mating Systems

Thus, another general consideration related to body size that might influence the relative timing of sex change across species is the mating system, the basic notion being that when larger males can monopolize matings with females, male size and male fitness tend to be strongly correlated, thus favoring sex change to maleness in larger and older individuals. Conversely, in some species such as the polychaete worm *Ophryotrocha puerilis*, male reproductive success drops with increasing size because, due to costly

FIGURE 3.16   *Gnorimosphaeroma oregonense*, a protogynous intertidal isopod.

conflicts over sex, females often prefer to mate with smaller males (Premoli and Sella 1995). Not surprisingly then, this species tends toward protandry (or at least toward protandric simultaneous hermaphroditism). More generally, in sequentially hermaphroditic animals for which the mating system is known, protogyny often tends to be associated with polygynous matings systems, whereas protandry often tends to be associated either with monogamous mating systems or with other situations in which males gain no substantial mating benefit from larger size (Munday et al. 2006). This statement holds especially for hermaphroditic fish (chapter 4), but several reports from invertebrate species are consistent with the notion as well (e.g., Collin 1995).

Conclusions About Sex Allocation Theory

Sex allocation theory (SAT) has been considered "a touchstone in the study of adaptation" (Frank 2002) and a highly successful body of evolutionary theory for understanding the adaptive significance of alternative reproductive modes (West et al. 2000). Studies on invertebrate animals have contributed to these glowing sentiments, and several of the predictions of SAT have received at least modest support. These include the link between small mating-group size and female-biased SA, and several observations about how SA varies with body size. But it is also true that several of the basic assumptions and predictions of SAT remain poorly supported by the current published literature on invertebrate animals. These include the assumed fundamental fitness trade-off between male and female allocation, and several more specific hypotheses such as that the phenomenon of brooding inevitably places strict limits on the reproductive returns from female allocation.

In a comprehensive literature review, Schärer (2009) emphasized the need for more experimental tests as well as observational studies of SAT in hermaphroditic invertebrates. There will also be a need to reconcile some apparent conflicts or paradoxes (Leonard 2005) between competing theories. For example, whereas SAT at face value (based on Bateman's principle) generally predicts that simultaneous hermaphrodites should be more eager to mate as males than as females because of the greater potential for increased fitness (Charnov 1979), precisely the opposite argument has been based on Gillespie's principle (Gillespie 1977) that hermaphrodites will prefer the female sexual role because its lower reproductive variance should mean a reduced probability of falling into the zero fitness class (Leonard 1999, 2006). These two viewpoints emphasize different approaches that organisms might take to the evolutionary game: striving to be a big winner in the reproductive sweepstakes (emphasizing the "upside potential") versus minimizing the probability of being eliminated from the contest entirely (focusing on the "downside potential").

In general, although predictions are rampant in the vast SAT literature, they often rest critically on assumptions that have been questionable or difficult to verify empirically. Thus, the relevance of SAT to empirical reproductive patterns in hermaphroditic invertebrates will require considerable clarification from further research.

Sexual Selection

As first appreciated and emphasized by Darwin (1871), the "sexual selection" that stems from intraspecific competition for mates can be a powerful driving force in evolution, often rivaling natural selection in terms of its phenotypic consequences. The concept of sexual selection can be applied at several biological levels including: the detection and attraction of mating partners; sperm competition and cryptic female choice; and even differential parental investment in progeny from different mates (Clutton-Brock 2004). Sexual selection clearly operates in many gonochoristic species, but can it impact hermaphroditic species as well? The short answer is a resounding "yes" (although this has not always been immediately evident).

For gonochoristic species, the most obvious sign of sexual selection is usually a pronounced phenotypic dimorphism between the sexes in traits that seem to be important in mate acquisition but are otherwise adaptively useless or even counterproductive (Shuster and Wade 2003). Classic examples in

TABLE 3.2   Several lines of circumstantial (and sometimes direct) evidence support the contention that sexual selection occurs routinely in hermaphroditic invertebrates (based on Leonard 2006)

Trait	Examples of Documentation	Reference or Review
Competition for mates	male fighting in polychaetes	Sella & Ramella 1999
Sperm competition	mollusks, polychaetes	Rogers & Chase 2001; Baur 1998; Michiels 1998; Lorenzi et al. 2006
Mate choice	numerous invertebrate taxa	Vreys & Michiels 1997; Leonard 2006
Sexual dimorphism	sea slugs and snails	Leonard & Lukowiak 1985; Ohbayashi-Hodoki et al. 2004
Elaborate or costly courtship	many gastropods	Baur 1998; Koene 2006; Michiels & Koene 2006
Special courtship structures	earthworms, gastropods	Adamo & Chase 1988; Koene et al. 2002
Conditional mating reciprocity	cestodes and snails	Milinski 2006; Webster & Gower 2006
Rapid genital evolution	gastropods, flatworms	Eberhard 1985
Skewed operational sex ratios	tapeworms	Schärer et al. 2001
Unequal variances in fitness for the two sexual roles	damselflies	Andersson 1994

vertebrate animals include a stag's antlers (used in male-male competition for mates) and a peacock's tail (which is cumbersome, but which peahens find attractive). In simultaneous hermaphrodites, however, the two sexes are combined in a single individual, so gender dimorphism in secondary sexual traits may not be so apparent. Indeed, simultaneously hermaphroditic species make it abundantly clear that phenotypic sexual dimorphism *per se* is not the ultimate hallmark or final arbiter of sexual selection.

However, several other phenomena that traditionally have been associated with sexual selection are indeed both commonplace and highly developed in hermaphroditic invertebrates and other dual-sex animals (Leonard 2006). These include: (a) costly, elaborate, and often bizarre courtship and copulatory behaviors (as in various snails, slugs, earthworms, and flatworms described earlier); (b) multiple mating, sperm competition, and probably cryptic female choice; (c) rapid evolution of genitalia in some lineages; (d) special morphological structures associated with courtship and mating; and (e) pronounced polymorphisms in sexual phenotypes and behaviors (as in phally-polymorphic snails). Furthermore, in some sequential hermaphrodites, secondary sexual characters in male and female phases of the life cycle show phenotypic differences analogous to those that distinguish males from females in many gonochoristic taxa. Several of these and other lines of empirical evidence for sexual selection in hermaphrodites are summarized in table 3.2.

The realization that sexual selection operates routinely in hermaphroditic as well as gonochoristic species should come as no surprise. In hermaphrodites and gonochorists alike, individuals struggle for genetic representation in subsequent generations. Although male and female gametes are housed in different ways (jointly in hermaphroditic individuals, separately in gonochorists), the never-ending fitness challenge of uniting one's own gametes with those from other individuals virtually ensures that sexual selection remains a major shaping force in the evolution of essentially all organisms that engage in sexual reproduction.

SUMMARY

1. Hermaphroditism is widespread and common in invertebrates, having been documented in more than 65,000 species representing nearly 70% of more than 30 taxonomic phyla. If the species-rich insects (none of which appears to be hermaphroditic) are excluded from consideration, then approximately 30% of the remaining invertebrate species are hermaphroditic. Simultaneous hermaphroditism is most prevalent, but many cases of sequential hermaphroditism also are known. Only a relatively few species, however, are androdioecious or gynodioecious.

2. The reproductive lifestyles of hermaphroditic invertebrates are incredibly varied and sometimes downright bizarre. Fascinating natural-history examples include the following: reef-building corals that spew gametes into the water much as wind-pollinated plants spew pollen into the air; terrestrial snails that shoot calcareous "love darts" into their mates during courtship and thereby deliver allohormones that manipulate sperm utilization by the recipient; freshwater snails in which the female part of the reproductive tract can store and use viable sperm for up to several months following a mating event; limpets that change sexual function from male to female as they age, and isopods that do exactly the reverse (switch from female to male) as they get bigger; polychaete worms that switch back and forth repeatedly between male and female; and slugs that engage in aerial courtship dances and exchange sperm while dangling on the ends of foot-long mucus strands.

3. In sharp contrast to the situation in plants, hermaphroditism is usually a derived condition in invertebrate lineages, both overall and in many taxonomic subclades. Exercises in phylogenetic character mapping (PCM) imply that hermaphroditism has arisen from gonochorism on many independent occasions, but also that many invertebrate taxa show evidence of phylogenetic constraints (and probably some evolutionary reversals as well). Intermediate evolutionary states have been harder to identify, however, in part because both androdioecy and gynodioecy are rare in extant invertebrate species and probably tend to be evolutionarily transient.

4. Several hypotheses have been advanced for why invertebrates commonly evolve dual sexuality from separate sexes. One long-standing notion is that hermaphroditism tends to evolve when female investment in ova is constrained for some reason, such as a paucity of brooding space or limited oviposition sites. Presumably, some resources then can be redirected to sperm production (ergo the emergence of hermaphroditism). Another and probably stronger hypothesis has to do with the fertilization insurance that hermaphroditism provides. A selfing hermaphrodite need encounter no one else to reproduce, and an outcrossing hermaphrodite need find just one other individual (as opposed to another individual of the proper sex, as is the case in gonochoristic species). Considerable circumstantial evidence is consistent with fertilization assurance being an important consideration in the emergence and maintenance of hermaphroditism.

5. Many species of simultaneous hermaphrodite display selfing as a component of a mixed-mating system. Regarding the fitness advantages and disadvantages of selfing vis-à-vis outcrossing, several considerations from genetics and ecology (plus interactions between the two) have been addressed. A key genetic consideration is inbreeding depression from continued selfing.

A key ecological consideration is the fertilization insurance that selfing provides. In some respects, a mixed-mating system may be a best-of-both-worlds outcome that combines the benefits of fertilization insurance from occasional selfing with the benefits of inbreeding avoidance via outcrossing.

6. Sex allocation theory (SAT) as applied to hermaphrodites traditionally assumes that all individuals in a population have a fixed or zero-sum budget for investments in male versus female reproductive functions. It seeks to understand how these resources are apportioned at any given time in simultaneous hermaphrodites, and when and why switches between investment tactics are accomplished in sequential hermaphrodites. Formal SAT has made many evolutionary predictions, such as: (a) increased selfing rates accompanied by increased allocations to female reproductive function; (b) an increased allocation to male reproductive functions for individuals in more stressful environments; (c) youthful reproduction as male when the initial costs of reproducing as a female are high or brooding constraints are severe; (d) youthful reproduction as female when older or bigger males can monopolize matings; (e) heavier investments in male function under conditions of intense local competition for mates or resources; and several others. Empirical tests of such predictions have produced mixed results, with some hypotheses—such as (a) and (d)—generally being well supported but others—such as (b) and parts of (c)—much less so.

7. For outcrossing simultaneous hermaphrodites, mating partners may not always be in lock-step with regard to preferred reproductive tactic. This can lead to sexual conflicts and to sexually antagonistic coevolution. In theory, sexual conflict can be resolved in various ways (such as conditional gamete trading), but no single class of theoretical models has as yet proved sufficient to explain all nuances in the empirical diversity of mating systems across hermaphroditic invertebrates.

8. Sexual selection (which stems from intraspecific competition for mates) applies with full force to hermaphroditic taxa. Empirical evidence for this statement comes from several sources, including the common appearance of: (a) costly and elaborate courtship and copulatory behaviors; (b) multiple mating, sperm competition, and probably cryptic female choice; (c) rapid evolution of genitalia; (d) special structures associated with courtship and mating; and (e) intraspecific polymorphisms in sexual phenotypes and behaviors. Individuals that simultaneously operate as both male and female have made it abundantly clear that phenotypic sexual dimorphism *per se* is not the ultimate hallmark or definition of sexual selection.

CHAPTER FOUR

Dual-sex Vertebrates

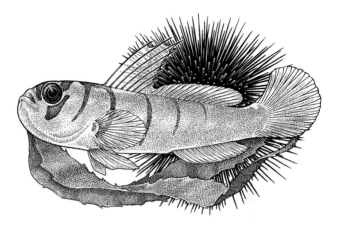

In vertebrate animals, true hermaphroditism (discussed further below) essentially is confined to fishes. This chapter will highlight piscine examples of operational hermaphroditism in both of its two primary reproductive forms: sequential (or serial) and simultaneous. Many of the conceptual and empirical topics that arose in discussions of hermaphroditic plants (chapter 2) and invertebrate animals (chapter 3) will again be germane to considerations of vertebrate species composed of dual-sex individuals.

The hermaphroditic condition has been described within more than 250 (ca. 1%) of the world's 25,000 extant species of teleosts or bony fishes (Pauly 2004). The phenomenon is represented by at least some species in approximately 30 taxonomic families distributed across nearly a dozen taxonomic orders. Most of these dual-sex fishes are sequential hermaphrodites and usually they are protogynous, but modest numbers of protandrous species, bi-directional sex changers, and simultaneous hermaphrodites are known as well.

Actually, characterizing whether or not a particular fish taxon is hermaphroditic is not always entirely straightforward (Sadovy and Shapiro 1987), especially because of an important distinction between the operational and the structural (anatomical) aspects of dual sexuality (Sadovy and Liu 2008). For example, occasional individuals in many fish species display various mixtures of both male and female body parts (including testicular and ovarian tissues) yet seldom if ever reproduce as functional hermaphrodites. Such individuals are sometimes referred to as intersexual fishes (box 4.1), and although they can be quite common in some otherwise gonochoristic species, we will not discuss them further in this chapter. Instead, we will confine

BOX 4.1   Intersexuality in Fishes

As was also true with respect to the phenomenon of intersexuality in *Homo sapiens* (see box 1.1), "intersexual" fish display various anomalies of the reproductive system (i.e., in the gonads or hormonal systems) that make particular individuals look "different" from standard males and females in an otherwise gonochoristic species.

For intersexual fish, the atypical sexual phenotype is normally observed only upon microscopic examination of internal gonadal tissue, which may display either of two types of anomaly: occasional female germ cells (oocytes) within a predominantly male gonad (Nolan et al. 2001); or occasional male germ cells (spermatocytes) within a predominantly female gonad (Vine et al. 2005). A recent report on the incidence of intersexuality in North American freshwater fishes (Hinck et al. 2009) showed the phenomenon to be surprisingly common and widespread: the intersex condition was present in approximately 3% of more than 3,000 individuals surveyed, in 25% of the 16 species examined, and at 31% of the 111 geographic sites sampled from across the continent.

The mechanism(s) responsible for intersexuality are incompletely understood but they probably include various exogenous (environmental) as well as endogenous (e.g., genetic, hormonal) influences in particular instances (Zillioux et al. 2001; Nash et al. 2004; Hecker et al. 2006). One major concern that has motivated studies on fish intersexuality is that environmental pollutants such as organochlorine pesticides, polychlorinated biphenyls (PCBs), heavy metals, and various pharmaceuticals might sometimes induce the intersex condition by disrupting the normal function of the piscine endocrine system.

"Abnormal hermaphroditism" is another topic closely related to the phenomenon of intersexuality. Atz (1964) defined "normal hermaphroditism" as a standard form of dual sexuality that exists, in a uniform way, at some time during the ontogeny of all or many members of a species that is genuinely hermaphroditic; all other forms of dual sexuality within an individual, including cases of intersexuality in otherwise gonochoristic taxa, would then be "abnormal hermaphroditism," by definition. Thus, although "hermaphroditism" occasionally has been reported in the scientific literature for various cartilaginous fishes including some sharks (e.g., Arthur 1950; King 1966; Yano and Tanaka 1989), most such instances probably entail abnormal hermaphroditism rather than normal hermaphroditism. Nevertheless, for at least one shark species (*Apristurus longicephalus*), a recent report of normal hermaphroditism also has appeared (Iglésias et al. 2005).

our attention to genuinely hermaphroditic taxa in which many or most "normal" individuals produce functional male and female gametes and thus have the standard potential to reproduce both as a male and as a female during their respective lifetimes.

Even so, the distinction between reproductive anatomy and reproductive function is important to bear in mind when considering the ecological or evolutionary ramifications of dual sexuality in fishes. For example, whereas a detailed comparison of gonadal structures might help to illuminate the phylogenetic relationships of various hermaphroditic taxa and the evolutionary histories of particular morphological features, a comparison of the fishes' actual reproductive modes would probably be even more important for understanding the selective pressures and the adaptive significance of alternative forms of hermaphroditic procreation.

Sexual Flexibility

The fact that many piscine lineages display functional hermaphroditism whereas other vertebrate lineages do not adds testimony to the general sexual flexibility of fishes vis-à-vis mammals, birds, and many reptiles and amphibians (Devlin and Nagahama 2002; DeWoody et al. 2010). This flexibility of fishes with respect to gender has two distinct but interrelated aspects: evolutionary and ontological.

Evolutionary Lability

Both mammals (class Mammalia) and birds (Aves) show a remarkable evolutionary conservatism of genetic sex-determining mechanisms. The familiar mammalian system of XY male heterogamety apparently arose in an ancestor of the therians (all extant mammals except the platypus) and has been retained by nearly all descendant mammalian lineages (Just et al. 1995; Lahn and Page 1999; Grützner et al. 2004; Waters et al. 2005; Potrzebowski et al. 2008). Similarly, a ZW sex-chromosome system of female heterogamety arose in an ancient avian ancestor and has been retained by nearly all extant birds (Fridolfsson et al. 1998; Mank and Ellegren 2007). In sharp contrast, mechanisms of genetic sex determination (GSD) are highly diverse in gonochoristic (separate-sex) fishes (box 4.2). Also, most of these GSD mechanisms are probably polyphyletic in fishes, each having evolved repeatedly and independently in various piscine lineages (Mank et al. 2006).

Furthermore, environmental sex determination (ESD) characterizes numerous fish taxa (Penman and Piferrer 2008), including most hermaphroditic species. Depending on the taxon, the environmental cues that elicit male versus female morphology and behavior at a particular stage of life

---

BOX 4.2   Genetic Sex Determination (GSD) in Gonochoristic Fishes

---

GSD mechanisms in one or another fish species are known to include the following: both XY (male-heterogametic) and ZW (female-heterogametic) sex-chromosome systems (Kikuchi et al. 2007; Vicari et al. 2008); combinations of competing male-heterogametic and female-heterogametic genetic mechanisms (Cnaani et al. 2007); multi-locus (polygenic) autosomal triggers (Vandeputte et al. 2007); interactions between sex chromosomes and autosomal modifier loci (Mair et al. 1991); and aneuploidy levels in sex chromosomes (Ribeiro de Oliveira et al. 2008; Ueno and Takai 2008). Approximately 10% of fish species have been documented to possess sex chromosomes (Devlin and Nagahama 2002), but the true incidence is probably closer to 50% when "nascent" sex chromosomes are taken into account (Arkhipchuk 1995). Nascent sex chromosomes are only partially differentiated from one another or from autosomes and, thus, are hard to detect in standard chromosomal assays. Many fish species without well-defined sex chromosomes nonetheless have GSD. This can happen when, for example, one or more sex-influencing genetic loci initiates a developmental cascade toward maleness, but genetic recombination near the gene(s) has not been suppressed to the extent that readily identifiable sex chromosomes have as yet emerged during the evolutionary process. For descriptions of specific genes known or suspected to be mechanistically involved in GSD in fishes, consult DeWoody and colleagues (2010).

Although GSD traditionally is distinguished from environmental sex determination (ESD) in fishes (see text), genetic and environmental factors sometimes interact to influence the sex of an individual. For example, sex determination in various tilapia species normally involves either the XY or ZW chromosomal systems, but the temperature at which embryos are reared can modulate or even override GSD in some cases (Mair et al. 1990; Desperz and Melard 1998; Baras et al. 2001). In the Atlantic silverside (*Menidia menidia*), different geographic populations differ in how they respond during development to the environmental cues that normally impact gender (Conover and Kynard 1981). More generally, modes of sex determination in fishes can range from pure GSD in some species to pure ESD in others, with some species showing various combinations of GSD and ESD.

---

may be physical factors such as water temperature (e.g., Conover and Heins 1987) and pH (Rubin 1985), or socio-demographic factors such as an individual's position in a dominance hierarchy or its behavioral interactions with potential mate(s) (Robertson 1972; Devlin and Nagahama 2002; Lorenzi et al. 2006; Rodgers et al. 2007; Baroiller et al. 2009; Mank and Avise 2009).

Thus, in fish species with ESD, the process of sex determination is endogenous or "genetic" only in the sense that the piscine genomes (and the developmental processes that they encode) have sufficient scope to permit phenotypically flexible gender responses to exogenous environmental influences (Harrington 1971; Craig et al. 1996; Fujioka 2001; Oldfield 2005). The phenomenon of ESD itself is also polyphyletic in fishes (Mank et al. 2006).

Ontogenetic Plasticity

In addition to its high permutability during piscine evolution, sex determination in fishes is highly flexible in an ontogenetic sense, and nowhere is this developmental plasticity more evident than in sequential hermaphrodites. In such species, by definition, at least some individuals switch from male to female (or vice versa) during their lifetimes. This phenotypic switch usually includes anatomical as well as physiological changes, and may be completed within a few days to several years depending on the species (Hattori 1991; Godwin 1994; Kobayashi et al. 2005). Extreme developmental flexibility with regard to gender in hermaphroditic fish has prompted extensive research into hormonal and other physiological aspects of the sexual transformations (reviewed by Francis 1992; Devlin and Nagahama 2002). However, the particular ontogenetic mechanisms need not concern us here; suffice it to say that testes and ovaries in teleosts both derive from the same precursor tissue and can rather flexibly differentiate at various life stages within an individual (Atz 1964). This contrasts with the situation in most other vertebrates (including all birds and mammals) in which gonadal differentiation activates early in development and generally is irreversible (Hoar 1969).

Sequential Hermaphroditism

In sequential or serial hermaphrodites, an individual produces eggs and sperm at different stages in life. Serial hermaphroditism is to be distinguished from simultaneous hermaphroditism, in which an individual produces male and female gametes at the same time.

The Cast of Players

The piscine orders, families, and approximate numbers of species in which one or another form of sequential hermaphroditism has been documented are listed in table 4.1, and a broad-picture phylogeny for these and other taxonomic orders is shown in figure 4.1. What follows are introductory descriptions of the natural histories of several of these taxa that have figured prominently in scientific investigations of sequential hermaphroditism in fishes.

TABLE 4.1 Taxonomic distribution of sequential hermaphroditism in teleost fishes, as compiled from the references indicated and from the additional reviews by Breder and Rosen (1966), Smith (1975), Mank (2006), and Mank et al. (2006). Families in bold font contain as least some protandrous species; italicized families contain at least some bi-directional sex changers; and all other families are primarily or exclusively protogynous.

Taxonomic Order	Taxonomic Family	Approximate No. of Hermaphroditic Species	Representative References
Anguilliformes	Muraenidae[4]	9	Devlin & Nagahama 2002
Clupeiformes	**Clupeidae**	2	Devlin & Nagahama 2002; Blaber et al. 1999
Cypriniformes	Cyprinidae	1	Devlin & Nagahama 2002
	Cobitidae	1	Devlin & Nagahama 2002
Cyprinodontiformes	Poeciliidae	1	Devlin & Nagahama 2002
Perciformes[1]	Callanthiidae	1	Devlin & Nagahama 2002
	Centropomidae	2	Taylor et al. 2000
	Centracanthidae	3	Atz 1964
	Cichlidae	1	Devlin & Nagahama 2002
	Cirrhitidae[6]	6	Devlin & Nagahama 2002; Kobayashi & Suzuki 1992
	Gobiidae[4]	>17	Devlin & Nagahama 2002; Francis 1992; Cole 1990; St. Mary 1998[2]
	Grammatidae	1	Devlin & Nagahama 2002
	Labridae	46	Devlin & Nagahama 2002; Francis 1992; Ebisawa 1990; Robertson 1972
	Lethrinidae	4	Devlin & Nagahama 2002
	Nemipteridae	7[5]	Devlin & Nagahama 2002
	Nototheniidae	1	Devlin & Nagahama 2002
	Pinguipedidae	1	Devlin & Nagahama 2002
	Polynemidae	1	Devlin & Nagahama 2002
	Pomacanthidae	19	Devlin & Nagahama 2002; Bruce 1980; Asoh & Yoshikawa 2003[3]
	Pomacentridae	9	Devlin & Nagahama 2002; Fricke & Fricke 1977

TABLE 4.1   (*continued*)

Taxonomic Order	Taxonomic Family	Approximate No. of Hermaphroditic Species	Representative References
	Pseudochromidae	1	Devlin & Nagahama 2002
	Scaridae	40	Devlin & Nagahama 2002; Choat & Robertson 1974
	Serranidae[4]	50	Devlin & Nagahama 2002; Atz 1964; Smith 1965; Brusle & Brusle 1974
	Sparidae	43[5]	Devlin & Nagahama 2002; Atz 1964; Francis 1992; Micale et al. 2002
	Terapontidae	2	Devlin & Nagahama 2002; Moiseeva et al. 2001
	Trichonotidae	1	Devlin & Nagahama 2002
Siluriformes	**Pimelodidae**	1	Quagio-Grassiotto & Carvalho 1999
Stomiiformes	**Gonostomatidae**	5	Devlin & Nagahama 2002
Synbranchiformes	Synbranchidae	2	Devlin & Nagahama 2002; Atz 1964

[1]As traditionally defined; however, Perciformes is probably polyphyletic.

[2]See also Robertson & Justines 1982; Munday et al. 2002.

[3]See also Moyer & Nakazono 1978.

[4]Some species in this taxon also are simultaneous hermaphrodites.

[5]A few of these species show only rudimentary structural hermaphroditism but functional gonochorism (Young & Martin 1985; Buxton & Clarke 1991; Garratt 1991; Lau & Sadovy 2001).

[6]Several of these species show an intermediate phenomenon between simultaneous and sequential hermaphroditism.

Groupers, Sea Basses, and Their Allies, Family Serranidae. Some of the largest of these marine fishes are as well known in culinary circles for their excellent taste as they are in scientific circles for their tendencies for protogynous hermaphroditism. Consider, for example, the Gag Grouper (*Mycteroperca microlepis*) (fig. 4.2), which is avidly sought by commercial and recreational fishermen in shallow coastal waters of the southeastern United States. Individuals in this species usually begin their reproductive careers

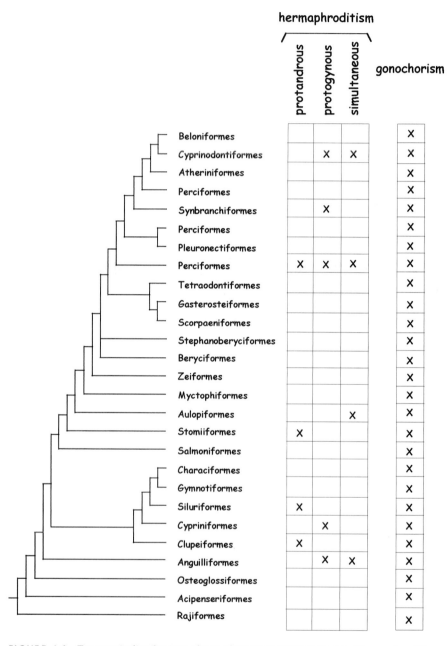

FIGURE 4.1 Taxonomic distributions of gonochorism and various forms of hermaphroditism along extant nodes of a provisional supertree for bony fishes (from Avise and Mank 2009, following Mank et al. 2006). The supertree is a cladogram (branch lengths are not proportional to divergence times) representing an amalgamation of several published phylogenies from whole-genome or partial-genome mtDNA sequences. Note that Perciformes is polyphyletic.

FIGURE 4.2   The Gag Grouper, a well-known protogynous hermaphrodite.

as egg-laying females, but many of them later in life change into sperm-producing males. Thus, males in this species tend to be older and larger (and also more aggressive) than females, and this has created some unanticipated difficulties for the fishery. On spawning reefs, belligerent males are especially vulnerable to baited hooks. As a consequence, males are caught disproportionately and have become rather rare (Chapman et al. 1999). For example, at one major spawning site off the east coast of Florida, the proportion of males has plummeted in the past two decades—to about 3% of the grouper population, compared to more than 10% before the fishery was overexploited.

A different expression of protogyny is provided by the Red Anthias, *Anthias* (or *Pseudanthias*) *squamipinnis* (fig. 4.3), which is an abundant serranid on coral reefs in the Indo-Pacific region. In this species, inch-long individuals form dense aggregations in which females may outnumber males by 36:1. If a small number (N) of males die or are removed, exactly N females may change sex to replace them. Such sex changes can also be triggered if the female:male ratio exceeds some threshold value (Shapiro 1979).

Wrasses, in the Family Labridae. This is another group of marine fishes in which protogynous hermaphroditism is common. Robertson (1972) provided

FIGURE 4.3   The Red Anthias, another protogynous hermaphrodite.

a classic description of one such species, the Indo-Pacific Cleaner Wrasse, *Labroides dimidiatus* (fig. 4.4). This species typically forms harems that consist of one large male and as many as ten females. Breeding access to the male is determined by a dominance hierarchy in which the largest female is dominant to the next-largest female, and so on down to the bottom of the pecking order. If the most dominant female dies or is removed, the next largest female assumes her role, and every other female likewise moves up in the social order. If the dominant male dies or is removed, the former alpha female begins courting females within an hour, and develops functional testes within two weeks.

The Caribbean Bluehead Wrasse, *Thalassoma bifasciatum* (fig. 4.5), displays a somewhat different expression of protogyny. Each fish normally begins life as a yellow female or as a similarly colored male in its "initial-phase" garb. Any of these yellow individuals may later change into a larger "terminal-phase" male who sports a bright blue head, a black-and-white midbody saddle, and a green posterior section. These large males then set up territories over coral heads that females utilize as spawning sites. Although

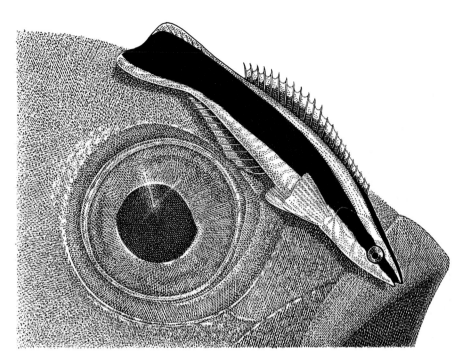

FIGURE 4.4   The Indo-Pacific Cleaner Wrasse, a haremic protogynous hermaphrodite.

smaller males occasionally spawn, the large alpha males are far more successful, sometimes spawning as many as 4–100 times per day (Warner et al.
1975; Warner 1991).

Porgies and Seabreams, Family Sparidae. This is a diverse group of more
than 100 marine species, many of which are hermaphroditic. Individuals in
some species have male and female gonads simultaneously but individuals
in other species may change sex as they grow. An example of the latter is the
Black Porgy, *Acanthopagrus schlegeli* (fig. 4.6), in which an individual typically functions as a male for the first two years of life before switching to
become a female.

Damselfishes, Family Pomacentridae. This family contains some of the best-
known representatives of protandrous hermaphroditism, classic cases-in-point
being clownfishes or anemonefishes such as the Clark's Anemonefish, *Amphiprion clarkii* (fig. 4.7). A clownfish was the famous protagonist in the Dis-
ney film *Finding Nemo* (although it did not show its hermaphroditic nature in
the movie). In real life, clownfishes inhabit coral reefs of the Indo-Pacific
where groups typically composed of two large individuals plus several small
specimens live together on a sea anemone. Only the two big individuals are
sexually mature, the largest being female and the other being male. If the

FIGURE 4.5 The Caribbean Bluehead Wrasse, another protogynous hermaphrodite. Shown are both the initial-phase and terminal-phase morphs.

FIGURE 4.6 The Black Porgy, a protandrous hermaphrodite.

FIGURE 4.7   The Clark's Anemonefish, a protandrous hermaphrodite.

female dies, the male changes sex to female and the next largest fish on the anemone takes over the role of the former male (Allen 1975; Fricke and Fricke 1977; Moyer and Nakazono 1978).

Bristlemouths, Family Gonostomatidae. These deep-sea (bathypelagic) fishes, an example being the Showy Bristlemouth (*Cyclothone signata*) (fig. 4.8), provide additional examples of protandry. An individual typically matures first as a male and then later may switch sex to become a female. The functional males generally remain smaller as adults and have a better-developed sense of olfaction, presumably to help them find females in the vast blackness of the deep ocean.

Gobies, Family Gobiidae. In the Redhead Goby, *Paragobiodon echinocephalus* (fig. 4.9), as well as in several other coral-dwelling goby species in the genera *Gobiodon* and *Paragobiodon*, an individual may change back and forth serially between male and female sexual roles during its lifetime (Sunobe and Nakazono 1993; Kuwamura et al. 1994; St. Mary 1994, 1997). These reef-inhabiting fishes typically live among the protective branches of live corals, often as single breeding pairs but sometimes in larger social groups in which only the biggest two or three individuals are reproductively active.

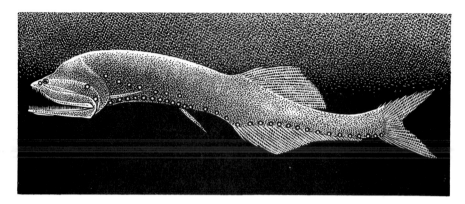

FIGURE 4.8   The Showy Bristlemouth, another example of a protandrous hermaphrodite.

FIGURE 4.9   The Redhead Goby, one of many goby species with serial bi-directional sex change.

Evolutionary History

The taxonomic orders of bony fishes in which hermaphroditism has been documented are scattered widely across the phylogeny of teleosts (see fig. 4.1). No order consists solely of hermaphroditic species, but each instead contains gonochoristic species also. Indeed, closer phylogenetic inspection

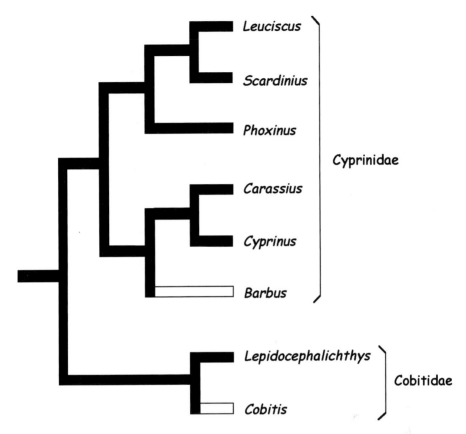

FIGURE 4.10 Phylogenetic position of sequential hermaphroditism (white branches) within cypriniform fishes (after Mank et al. 2006). Black branches indicate gonochorism.

of several orders and families that are polymorphic for reproductive mode indicates that each hermaphroditic clade typically is embedded in a deeper clade that is otherwise composed of gonochoristic taxa (examples are shown in figs. 4.10 and 4.11).

Extant hermaphroditism in teleost fish is therefore a polyphyletic and derived condition relative to gonochorism. Furthermore, no extant hermaphroditic lineage appears to be evolutionarily ancient. The evolutionary flexibility (and probable polyphyly) of reproductive modes in fishes also extends to the major forms of sequential hermaphroditism: protogyny and protandry (fig. 4.12).

How exactly does hermaphroditism evolve from gonochorism in fishes? And does gonochorism ever reemerge during evolution from phylogenetically localized instances of ancestral hermaphroditism? Definitive answers to

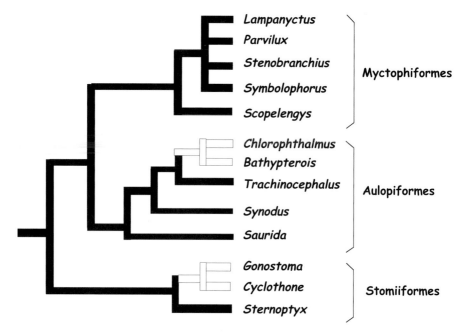

FIGURE 4.11   Phylogenetic position of hermaphroditism (white branches) within aulopiform and stomiiform fishes (after Mank et al. 2006). Black branches indicate gonochorism.

such questions are generally lacking for fishes. However, perhaps some conceptual insight can be gleaned from the botanical literature (chapter 2), where theoretical and empirical methods have been used to good effect to elucidate the evolutionary history of hermaphroditism in plants. For plants and fishes alike, in principle, gonochorism and hermaphroditism are on opposite evolutionary poles of a reproductive spectrum that includes various intermediate mixed-sex modalities (fig. 4.13): gynodioecy (a population mixture of females and hermaphrodites), androdioecy (a population mixture of males and hermaphrodites), and/or trioecy (a population mixture of males, females, and hermaphrodites). Trioecy is extremely rare in the biological world and androdioecy is only slightly less so (Weeks et al. 2006), but gynodioecy is rather common (especially in plants, where >500 species in 50 families display the phenomenon; Jacobs and Wade 2003). As detailed in chapter 2, theoretical models addressing the evolution of these mixed-sex reproductive systems in plants have a long history (Lloyd 1975; Charlesworth and Charlesworth 1979; Charnov 1982; Charlesworth 1984; Jarne and Charlesworth 1993; Barrett 1998; Pannell 2002; Wolf and Takebayashi 2004).

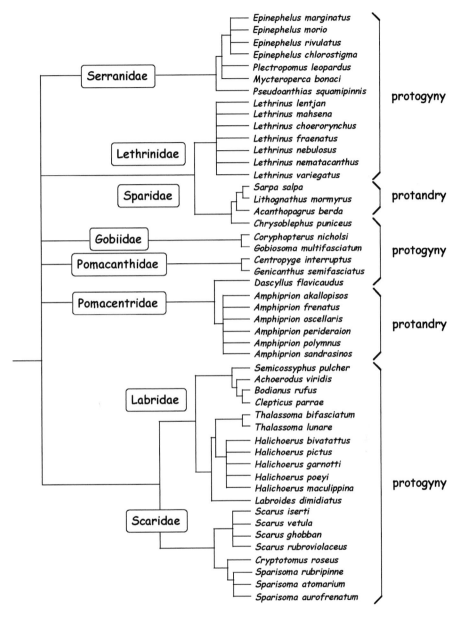

FIGURE 4.12 Phylogenetic interspersion of protogyny and protandry in Perciformes (after Allsop and West 2003). The diagram shows only hermaphroditic species; many other species in this traditional taxonomic order (which itself is polyphyletic, with family relationships often poorly known) are gonochoristic.

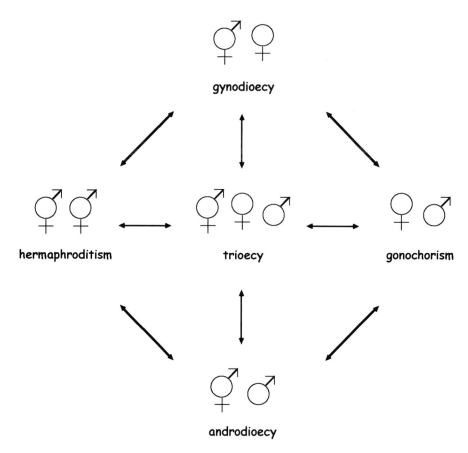

FIGURE 4.13   Reproductive systems that in principle could be evolutionarily transitional between gonochorism and hermaphroditism, e.g., in plants (modified from Weeks et al. 2006).

From this botanical literature, the following points have emerged. During an evolutionary transition between separate-sex and purely hermaphroditic reproduction, typically one sex at a time is either lost (in a transition from gonochorism to hermaphroditism) or gained (in a transition from hermaphroditism to gonochorism). This factor alone (a step-by-step transition) probably helps to account for the rarity of trioecy. Furthermore, in any evolutionary transition from gonochorism to hermaphroditism (or vice versa), gynodioecy has been deemed theoretically more likely than androdioecy because, ultimately, male gametes are far more abundant than female gametes. Thus, especially if hermaphrodites can self-fertilize, with regard to reproductive success a pure female in a gynodioecious population should be more competitive than a male in an androdioecious population, assuming that male gametes from hermaphrodites can be used to fertilize a female's eggs. Once females have arisen in an otherwise hermaphroditic species, they may also gain a reproductive benefit by virtue of engaging in out-crosses only, whereas some lineages of self-fertilizing hermaphrodites might suffer from inbreeding depression.

Most of the theoretical models underlying these conclusions have in-corporated biological assumptions that apply to plants (and perhaps to some invertebrate animals) but may have less relevance to fish. For example, the assumption that male gametes from hermaphrodites are physically available to fertilize eggs of pure females may hold for many wind-pollinated plants and free-spawning marine invertebrates, but it may be inappropriate for fish species with elaborate courtship and spawning rituals; and self-fertilization is unknown in fishes, except in one androdioecious clade (*Kryptolebias*, to be discussed later in this chapter). Furthermore, some of the biological phe-nomena in plants (such as the prevalence of gynodioecy over androdioecy) that motivated the available models on evolutionary transitions between dioecy and hermaphroditism do not seem to apply to fishes. Indeed the conditions of gynodioecy and androdioecy are only rarely reported in fish (Robertson et al. 1982; Petersen and Fischer, 1986).

Another reservation about applying the theory developed for plants to fishes involves the implicit assumption that the evolutionary states transi-tional to gonochorism and hermaphroditism are genetically hardwired and thus directly responsive to natural or sexual selection. Instead, sexual dif-ferentiation in fish is remarkably plastic developmentally, and subject to en-vironmental influences (Francis 1992). For example, in the androdioecious Mangrove Rivulus, males can be experimentally induced by exposure to particular environmental conditions (Harrington and Kallman 1967); and in several sequentially hermaphroditic fishes, developmental switches be-tween male and female have long been known to be socially mediated (Fish-elson 1970; Robertson 1972; Fricke and Fricke 1977; Shapiro 1979). As already mentioned, these broad norms of reaction with respect to gender reflect the fact that testes and ovaries in teleosts derive during ontogeny from a single precursor tissue that can differentiate rather flexibly during an individual's lifetime. This is not to imply that selection plays no role in the evolution of alternative reproductive modes in fish, or that fish reproductive operations have no genetic basis. To the contrary, proximate environmental factors that influence sexual expression in a fish population are likely to alter the selec-tion pressures that ultimately influence the evolution of underlying sex-influencing mechanisms (including, in some species, the genetic and devel-opmental scope for hermaphroditism).

In any event, the apparent paucity of gynodioecy and androdioecy in extant fishes suggests that these transitional states tend to be evolutionarily ephemeral and hence rare at best, perhaps because the set of fitness conditions favoring the stability of mixtures of hermaphroditic and gonochoristic systems is restrictive (Charnov et al. 1976). In part for this reason, most of the available evolutionary theory regarding hermaphroditism in fishes has focused not so much on the transitional states leading to hermaphroditism, but instead on the selective factors that might promote the expression of different forms of hermaphroditism in various taxa.

Adaptive Significance of Alternative Modes

Earlier researchers sometimes invoked population-level advantages (i.e., "group selection") to rationalize the evolution of hermaphroditism. For example, Moe (1969) suggested that sequential hermaphroditism might have evolved as a population control mechanism, with the age of transformation between female and male shifting up or down to compensate for whether a population was too spare or too dense in a particular sex. Other hypotheses with a group selection aura posited that hermaphroditism might increase total zygotic production in a population (Smith 1967) and/or focus sexual performances into age classes that would maximize a population's reproductive output (Nikolski 1963).

Furthermore, some ichthyologists of past decades assumed that each sequential hermaphrodite automatically changed sex upon reaching a threshold body size or critical age. However, laboratory and field observations have demonstrated that social and behavioral factors typically trigger each switch from one sex to the other, and that this can happen at different ages or sizes of the fish involved (Shapiro 1987). For example, removal of a dominant male from a social group of protogynous fish, or removal of a female from a protandric group, may induce one or more remaining individuals to change sex. Nevertheless, at least as an empirical generality, Allsop and West (2003) noted that hermaphroditic fish tend to change sex when they reach about 80% of their maximum body size and are about 2.5 times their initial age at sexual maturity. Considerable effort has gone into analyzing possible ecological and demographic conditions that proximately trigger such sex changes and that ultimately have led to protogyny, protandry, or serial sex-switching in particular fish species that currently display one or another of these alternative expressions of sequential hermaphroditism.

In sharp contrast to some of the earlier group-selectionist notions mentioned above, modern views have emphasized how natural selection and sexual selection might operate, in various social and ecological contexts, on the *differential fitnesses of individuals* who have the capacity to reproduce as males *and* as females at various stages of life (Warner 1975; Shapiro 1987; Leonard 2006). One of the first of these modern evolutionary models for sex change—the "size-advantage" hypothesis—today remains a singularly powerful explanation for the various patterns of sequential hermaphroditism observed in fishes.

As originally phrased by Ghiselin (1969:190), "Suppose that the reproductive functions of one sex were better discharged by a small animal, or those of the other sex by a large one. An animal which, as it grew, assumed the sex advantageous to its current size would thereby increase its reproductive potential." The basic approach of the size-advantage model is to plot (as in fig. 4.14) expected age-specific or size-specific fecundity under biological assumptions presumably relevant to the population under consideration. If

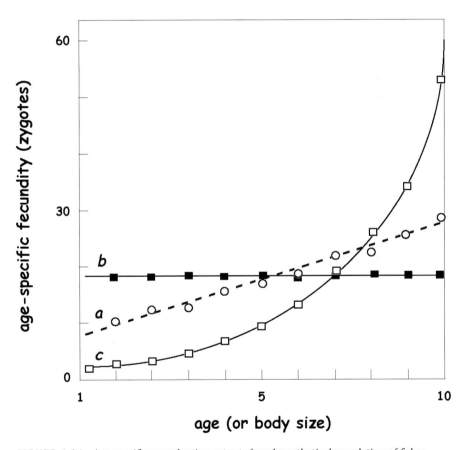

FIGURE 4.14   Age-specific reproductive outputs for a hypothetical population of fishes (from Avise and Mank 2009, after Warner 1975). Curve *a* is for females; curve *b* is for males in a population where mating is random; and curve *c* is for males in a population where females mate only with same-age or older males (as might tend to be true in male-territorial or haremic species, for example; but see also Taborsky 2008).

the curves relating fecundity to age cross for the two sexes, then, in principle, an age or body size exists at which an individual could reproductively profit by switching gender (see fig. 4.15). This life-history notion was formalized by Warner (1975) and Warner and colleagues (1975), who showed that if age-specific reproductive output increases more rapidly with age (or body size) for one sex than the other, then "individuals should change sex when the reproductive prospects of functioning as the opposite sex exceed the expectations of the current sex" (Warner and Swearer 1991:204).

This size-advantage or age-advantage model has stimulated a large and successful body of empirical as well as theoretical research on hermaphroditism in fishes. As noted by Shapiro (1987:44): "If we assume that the mechanism controlling induction of sex change has evolved to permit individuals to change sex only when it is to their reproductive advantage to do so, then a satisfying, evolutionary explanation should be capable of predicting correctly when individuals should change sex and when they should not."

Following the seminal treatments that introduced the size-advantage model, most subsequent empirical appraisals and theoretical analyses of sequential hermaphroditism can be considered refinements that have taken into account additional factors (beyond body size *per se*) that might impact age-specific fecundity and mortality curves in ways that affect individuals'

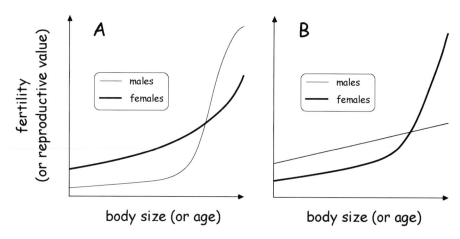

FIGURE 4.15　The size-advantage hypothesis (after Munday et al. 2006). A: expected production of offspring by males increases faster than that of females at larger body sizes. B: expected production of offspring by females increases faster than that of males at larger body sizes. Sex change is favored when the size-specific or age-specific fertility curves of the two sexes cross, such that protogyny is favored in the situation illustrated by panel A and protandry is favored in the situation illustrated by panel B.

expectations for reproductive success as a function of gender. The kinds of complicating (and often interacting) factors that have been addressed include population density (Warner and Hoffman 1980; Lutnesky 1994), population body-size ratios (Ross et al. 1983), sex ratios and mating patterns, immediate physiologic or other costs (including missed mating opportunities) of sex change *per se* (Hoffman et al. 1985; Iwasa 1991; Munday and Molony 2002), or other life-history trade-offs (Charnov 1996). Each such factor has been considered in the context of how it might alter age-specific reproduction (and, hence, selection pressures for the timing and direction of sex change) in one species or another of sequentially hermaphroditic fish.

Protogyny (female-first sequential hermaphroditism) is the most common form of dual sexuality in fishes; it characterizes, for example, many wrasses (Warner and Robertson 1978), parrotfishes (Robertson and Warner 1978), groupers, sea basses, and other reef fish. The Bluehead Wrasse (*Thalassoma bifasciatum*) provides a well-studied example (Warner and Swearer 1991). As mentioned earlier, females and juvenile males display an initial phase (IP) coloration with a yellow dorsal stripe and a series of lateral green blotches separated by white bars, whereas large breeding males show a striking terminal phase (TP) with a bright blue head and green body. The IP males are "primary" males, whereas those in the nonreversible terminal phase are "secondary" males that arose either from IP males or from particular females who changed sex (typically within a few days following the loss of TP males from a local population). Some other protogynous species have roughly similar lifestyles but are monandric: i.e., essentially *all* males derive from sex-changed females. One such apparent example is the Potter Angelfish, *Centropyge potteri* (fig. 4.16).

Protogyny is predicted to be evolutionarily favored when the reproductive output of males increases, as a function of size or age, faster than that of females. Thus, protogyny should often be associated also with sexual selection operating on males. Male size advantage is especially likely when large males tend to monopolize matings (Warner 1988), as for example in species where territorial or otherwise ruling males control reproductive access to females (Ross 1990). Thus, it is probably no mere coincidence that protogynous life histories (and pronounced sexual dimorphism) are observed most frequently in fishes with haremic social systems (Lutnesky 1994) in which most of the mating events are instrumented by large, dominant males (although it may be difficult in particular cases to establish whether protogyny is the cause or the effect in this association).

Although the size-advantage model generally has proved powerful in explaining protogynous sex change in hermaphroditic fish species, not all field observations seem easily accommodated under this model. For example, in some protogynous species the largest females do not always change sex when given the opportunity (Lutnesky 1994; Cole and Shapiro 1995). To

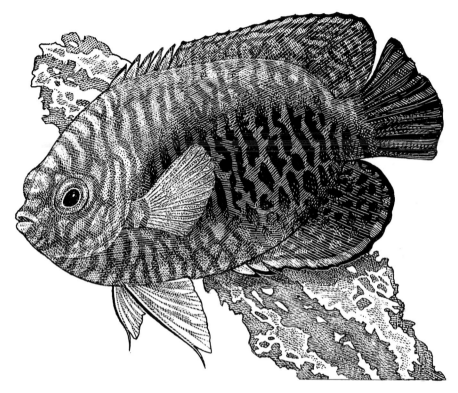

FIGURE 4.16   Potter's Angelfish, a species in which all males formerly were females.

address this conundrum, Muñoz and Warner (2004) conducted field observations and experiments on a Caribbean population of Bucktooth Parrotfish, *Sparisoma radians* (a relative of the Queen Parrotfish pictured in figure 4.17). The authors found that pronounced size-related skews in female fecundity, coupled with dilutions of paternity via inter-male sperm competition, set up population conditions in which, for the largest females, expected reproductive success as a male was actually lower than continued reproduction as a female. Factoring these complications into the models helped to make sense of the observation that smaller females were often the sex changers in this species. This study illustrates how suitable modifications to the size-advantage model have sometimes proved useful in understanding the peculiarities of particular protogynous systems.

Protandry (male-first sequential hermaphroditism) is a somewhat less common mode of dual sexuality in marine fishes, but the phenomenon is known for example in various anemonefishes (Pomacentridae; Miura et al. 2003),

FIGURE 4.17   The Queen Parrotfish, *Scarus vetula*, which like many of its cousins is a protogynous hermaphrodite.

porgies (Sparidae; Wu et al. 2005), and snooks (Centropomidae; Taylor et al. 2000). In the case of anemonefish, a local breeding community typically consists of one dominant female and several smaller males and juveniles, and if the breeding female dies a male then transforms to take her place.

Perhaps the most surprising aspect of protandry is its rarity relative to protogyny. In most fish species, female fecundity increases dramatically with age and body size, whereas even small mature males can produce enough sperm to fertilize countless eggs. Thus, at least at face value, selection pressures might generally seem to favor individuals who can adopt a male-first-in-life strategy (all else being equal). On the other hand, protandry typically produces a male-biased sex ratio, thereby exacerbating the sperm excess that is expected even in populations with a more balanced sex ratio. Thus, especially where intense male competition exists for mating opportunities, protandry might be generally counterproductive to individual fitness. Additionally, for many fish species in which mating is either random or monogamous, the slopes in the regressions of age-specific fecundity on body size are probably rather similar in males and females (compared to the great disparity in these slopes, especially in later age cohorts, when large males are highly polygynous and can monopolize matings with many females). Thus, by these latter lines of reasoning, the rarity of protandry relative to protogyny might not be so unexpected after all.

Although individuals in most sequentially hermaphroditic fish species change sex only once in a lifetime, in several goby species (Gobiidae) individuals routinely display serial sex changes in both directions (Sunobe and Nakazono 1993; Kuwamura et al. 1994; St. Mary 1994). Most notable among these gobiid species are obligate coral-dwellers that live among the protective branches of live corals, often as breeding pairs but sometimes in larger social groups in which only the biggest individuals normally reproduce.

Two evolutionary hypotheses have been advanced to account for bidirectional sex change in such species. Under the "risk-of-movement" model (Nakashima et al. 1996; Munday et al. 1998), intense predation pressures on patch-structured reefs make mate-searching movements very risky for small and sparsely distributed fish like gobies, thus giving a selective advantage to any stay-at-home individuals who can facultatively switch gender as the need arises (such as when a mate dies or when the sex ratio is highly skewed in the local environs). A different (but somewhat overlapping) hypothesis—the "growth-rate-advantage" model (Kuwamura et al. 1993; Nakashima et al. 1995)—incorporates the observation that female gobies grow faster than males, yet reproductive success may increase equally with body size in both sexes (because larger females produce more eggs and larger males can better defend egg clutches). Under this biological setup, when two potential mates meet, selection should favor different kinds of sex change: protogyny when the initial pair consists of two females, protandry when the pair consists of two males, and sex reversal when the female initially is larger than the male in a heterosexual pair (Munday 2002). The two competing models—risk-of-movement and growth-rate-advantage—were put to test using manipulative field experiments for the Australian goby *Gobiodon histrio*, and for this species the risk-of-movement hypothesis proved to match the empirical observations most closely (Munday 2002).

Simultaneous Hermaphroditism

The Cast of Players

In a rather small number (ca. 40) of fish species, at least some individuals produce and deploy both male and female gametes simultaneously (Fischer and Petersen 1987; Cole 1990; Kobayashi and Suzuki 1992), i.e., within the same reproductive episode or season. "Dual sexuality" of this form is known in representatives of about half a dozen taxonomic families, most notably in the Serranidae, Cirrhitidae, and Gobiidae (table 4.2). As will be detailed later, nearly all synchronously hermaphroditic fishes outcross (mate with other individuals) exclusively, rather than self-fertilize.

TABLE 4.2   Taxonomic distribution of simultaneous hermaphroditism in teleost fishes, as compiled from the references indicated and from the additional reviews by Breder and Rosen (1966), Smith (1975), Mank (2006), and Mank et al. (2006).

Taxonomic Order	Taxonomic Family	Approximate no. of hermaphroditic Species	Representative References
Anguilliformes	Muraenidae[6]	3	Devlin & Nagahama 2002
Aulopiformes	Chloropthalmidae	1	Devlin & Nagahama 2002
	Ipnopidae	1	Fishelson and Galil 2001
Cyprinodontiformes	Rivulidae	2	Harrington 1971; Tatarenkov et al. 2009
Perciformes[1]	Cirrhitidae[2]	6	Kobayachi & Suzuki 1992; Devlin & Nakahama 2002
	Gobiidae[3]	10	Devlin & Nagahama 2002; Cole, 1990; St. Mary, 1993[4]
	Serranidae[6]	14	Devlin & Nagahama 2002; Atz 1964; Brusle & Brusle 1974; Francis 1992[5]

[1]As traditionally defined; however, Perciformes is probably polyphyletic.

[2]Several of these species show bi-directional sex changes and/or intermediate phenomena between simultaneous and sequential hermaphroditism.

[3]Several of these species tend toward protogynous hermaphroditism rather than full sexual simultaneity.

[4]See also Kuwamura & Nakashima 1998.

[5]See also Barlow 1975.

[6]Some species in this taxon also are sequential hermaphrodites.

Evolutionary Constraints

Like sequential hermaphroditism, synchronous hermaphroditism is undoubtedly a polyphyletic condition in fishes (see fig. 4.1), having arisen independently on at least several different occasions in distinct piscine lineages. The relative rarity of synchronous hermaphroditism in fish is probably related to difficulties in the evolutionary origin or maintenance of joint sexuality due to inherent antagonisms within an individual between male and female hormones or other physiological systems (Bull and Charnov 1985), and/or to high fixed costs associated with maintaining both sexual functions within an individual at the same time (Heath 1977). Additionally, for simultaneous hermaphrodites that outcross (and nearly all do), the high cost of producing female gametes creates for each individual an inherent and perhaps substantial fitness risk: without some mechanism to ensure that mates reciprocate by contributing proper amounts of both gametic types, cheating strategies might readily evolve and thereby lead to the decay or loss of simultaneous hermaphroditism within a lineage. This kind of evolutionary instability might have further contributed to the relative paucity of simultaneous hermaphroditism in extant fishes.

Mating Behaviors in Outcrossers

With only one known exception (to be described later), simultaneously hermaphroditic fish species reproduce exclusively via outcrossing (in which separate individuals mate), rather than by self-fertilization. In some serranid species such as the Chalk Bass, *Serranus tortugarum* (fig. 4.18), each

FIGURE 4.18   The Chalk Bass, *Serranus tortugarum.*

FIGURE 4.19   The Black Hamlet, *Hypoplectrus nigricans.*

hermaphroditic individual in a spawning pair typically alternates sexual roles in close succession, spawning serially during an encounter sometimes as a male and sometimes as a female; and in a variant of this behavioral theme in the Black Hamlet, *Hypoplectrus nigricans* (fig. 4.19), two otherwise solitary individuals pair up (typically in the late afternoon) to spawn in a process called egg trading that consists of a three-step behavioral sequence (Fischer and Petersen 1987): (a) each fish packages an entire day's clutch of eggs into parcels; (b) courtship is initiated by the individual that will first release eggs; and (c) partners take turns releasing an egg parcel every few minutes and externally fertilizing their mate's released parcel. Hamlet partners are usually faithful during the episode but they occasionally switch partners on different days, so the mating system approximates serial monogamy, albeit in a compressed timeframe. In the Harlequin Bass (*Serranus tigrinus*), the process is similar except that clutches are not parceled and the mating system is thought to be "permanent" monogamy.

In other species such as the Blue-banded Goby, *Lythrypnus dalli* (fig. 4.20), an individual has competent male and female gonadal tissue simultaneously but nonetheless normally acts only as a male or a female at each stage in life, thus partly decoupling the physiological and the behavioral aspects of synchronous hermaphroditism (St. Mary 1993). Thus, although such species are simultaneously hermaphroditic in an anatomical sense, they might better be considered sequentially hermaphroditic in a functional sense. Finally, in a

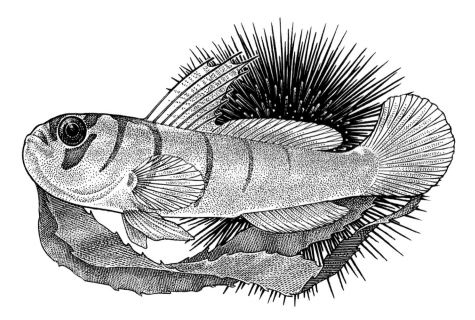

FIGURE 4.20   The Blue-banded Goby, *Lythrypnus dalli.*

few species including the Barred Serrano, *Serranus fasciatus* (fig. 4.21), most
specimens mature as simultaneous hermaphrodites but larger individuals
later may lose female function and become males (Petersen 1990). Thus, such
species could also be deemed androdioecious (Hastings and Petersen 1986).
The mating system of the Barred Serrano is a harem-style polygyny that is
quite similar to that of some of the protogynous wrasses that were described
earlier in this chapter.

In general, outcrossing synchronous hermaphrodites face sex-role decisions within a spawning episode or season that are analogous to the sex-role
choices faced by sequential hermaphrodites across a lifetime: namely, how
best to allocate male versus female function so as to maximize expectations
for current and future reproductive success (Fischer 1981, 1984; Petersen and
Fischer 1986; St. Mary 1994). Thus, sex allocation theory—which addresses
how a hermaphrodite should in principle divide its reproductive portfolio
between male and female efforts (Petersen 1990)—is again relevant, and
many similar considerations that arose for sequential hermaphrodites reap-
pear in this new context (Charnov 1996). For example, in the Barred Serrano—a species with female harems—a late-life switch from hermaphrodite to male can be rationalized by the size-advantage hypothesis, using
the same basic argument (disproportionate mating success for larger males)

FIGURE 4.21   The Barred Serrano, *Serranus fasciatus.*

that applied to haremic species of protogynous hermaphrodites (Petersen 1987).

With respect to egg-trading behavior in some of the serially monogamous serranids such as the Black Hamlet, a "tit-for-tat" model (box 4.3) has been applied (Fischer and Petersen 1987; Petersen 1995). The basic idea is that by releasing a clutch gradually and waiting for a partner to reciprocate, a pair-mating fish can better evaluate the mating situation and cut its losses if its partner deserts. In the Harlequin Bass, by contrast, egg parceling may be somewhat less critical because the monogamous pair bond tends to have a greater permanency. In general, tit-for-tat scenarios are merely one aspect of the "hermaphrodite's dilemma" (Leonard 1990) that envisions inevitable sexual conflict (differences of interest between male and female partners; Leonard 1993) in any reproductive interaction involving reciprocity with possible cheating. In turn, the interactive reproductive games played by hermaphroditic fish are just one subset of the much broader topic of how cooperative interactions might evolve in any animal species (Axelrod and Hamilton 1981).

Another basic adaptive consideration for synchronous hermaphrodites (but not for all sequential hermaphrodites) is encapsulated in the "low-density" model, which notes that individuals who produce male and female gametes at the same time should have less difficulty than gonochorists in finding mates, especially when populations are sparse (Tomlinson 1966). This fertilization advantage should hold both for outcrossing hermaphrodites

BOX 4.3   The Tit-for-Tat Model and the Hermaphrodite's Dilemma

Game theory (chapter 1) is often applied to situations in which the inter-actions between pairs of individuals (such as potential mating partners) impact the participants' personal payoffs (genetic fitness in this case). One common situation is known as the "prisoner's dilemma" in which each of two partners has two options: C, cooperate with your partner; or D, defect. The payoff matrix in this famous example of game theory is as follows:

		Partner	
		D	C
Focal	D	2	4
individual	C	1	3

Thus, under this setup, regardless of what his partner does, a focal indi-vidual who is rational should always defect because he thereby gets a greater reward than if he cooperates. But presumably the partner is rational too, so the expected payoff is 2. Ironically, the payoff would be higher (3) if both partners cooperated, thus creating a paradox. One escape from this para-dox is to suppose that the game is played repeatedly by the same two play-ers and that each player adopts a tit-for-tat (TFT) tactic with the following rule: cooperate in the first round and then in each subsequent round play the move that your partner played in the previous game. Formal game theory has shown that such a TFT approach can be an evolutionary stable strategy (ESS) once it evolves (Maynard Smith 1998:166–67).

For simultaneously hermaphroditic fishes in the family Serranidae, the TFT model has been invoked to explain the egg-trading phenomenon wherein spawning individuals often donate eggs to be fertilized in ex-change for the opportunity to fertilize the eggs of a partner (Fischer 1988). According to Fischer (1988), this form of delayed reciprocity is especially likely to originate and persist in the context of the monogamous mating systems of the sort that characterize several serranid species, because in these cases the same pair of individuals has multiple spawning interac-tions that can permit the TFT behavior to emerge as a potential resolution of the hermaphrodite's dilemma.

(who need to encounter only one other individual to mate) and for selfing hermaphrodites (who need not encounter any partner). In this mate-acquisition regard, some of the possible selective advantages for synchronous hermaphrodites can overlap those for sequential hermaphrodites under the risk-of-movement model described earlier.

Self-fertilization

Only one tiny evolutionary clade of hermaphroditic fishes—containing the Mangrove Rivulus (fig. 4.22), *Kryptolebias* (formerly *Rivulus*) *marmoratus*; Rivulidae—is documented to self-fertilize routinely. Most mature individuals have an internal ovotestis that produces sperm and eggs that typically unite inside a fish's body, after which zygotes are laid into the environment. Each hermaphrodite deposits its fertilized eggs in shallow water or on moist soil in the animal's coastal mangrove-forest habitat. Embryonic growth, hatching, and juvenile development then proceed without further parental involvement (Taylor 1990, 2000).

In some populations of *K. marmoratus*, selfing rates are so high that nearly all individuals belong to highly inbred lineages that have near-zero heterozygosities and thus, in effect, are clonal. Also present in *K. marmoratus* are pure males who appear to mediate occasional outcross events. This probably happens when a hermaphrodite sheds some unfertilized eggs onto which a male (who has no intromittent organ) releases sperm. The males are of two types: secondary males who are hermaphroditic when young but late in life lose

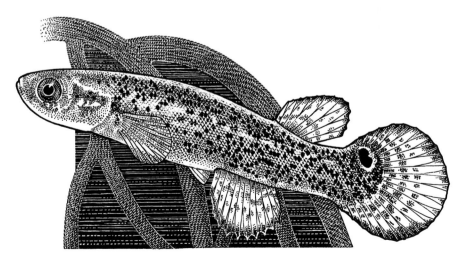

FIGURE 4.22   The Mangrove Rivulus, *Kryptolebias marmoratus*.

ovarian function (Harrington 1971); and primary males who have functional testicular but not ovarian tissue throughout life. Harrington (1967, 1968) discovered that he could readily generate primary males in the laboratory, for example by incubating self-fertilized eggs from a hermaphrodite at low temperature. However, *K. marmoratus* males seem to be very uncommon in eastern Florida (where Harrington and Kallman obtained their strains), and this observation in conjunction with the mostly clonal makeup of natural populations led the authors to conclude that outcrossing is rare or absent in nature (Kallman and Harrington 1964).

Males subsequently were uncovered in much higher frequencies (>20%) in some other populations of *K. marmoratus* (Davis et al. 1990), notably at Twin Cays in Belize (Turner et al. 1992a, 2006). Initial molecular assays (DNA fingerprinting) documented the occurrence of natural outcrossing at the Belize sites (Lubinski et al. 1995; Taylor et al. 2001), and recent genetic reappraisals based on highly polymorphic microsatellite loci suggest that about 55% of the matings in Belize may be outcross events and 45% involve selfing (Mackiewicz et al. 2006c). Outcrossing has also been genetically confirmed in Floridian populations (Mackiewicz et al. 2006b,c), albeit at much lower inferred frequencies (ranging from zero to 20% across ten surveyed locations). Outcross events are usually presumed to be male-mediated (Sakakura and Noakes 2000), because *K. marmoratus* hermaphrodites show courtship and spawning behaviors typical of females in other killifish species (Kristensen 1970) and because male participation has been documented in the laboratory (Mackiewicz et al. 2006a).

However, it remains possible that pairs of hermaphrodites sometimes cross as well. Another possibility is that some of the outcross events involve crosses between males and young individuals that function solely as females. Using gonadal dissections, Cole and Noakes (1997) found that some relatively young specimens of Mangrove Killifish are pure females that only later, in adult life, add sperm production to their overall reproductive repertoire. However, it remains unclear whether these young females actually reproduce, or whether they are merely in a transient developmental stage on the path to functional hermaphroditism. Regardless, such specimens raise the possibility that simultaneous hermaphroditism in *K. marmoratus* might have evolved from an intermediate condition of protogynous (female-first) hermaphroditism. In any event, *K. marmoratus* today can be described as an androdioecious species with a mixed-mating system of both selfing and outcrossing (Mackiewicz et al. 2006a,b, c).

The range of *K. marmoratus* extends from southern Florida to northern South America and includes at least some oceanic islands in the Caribbean. Overall, the genus *Kryptolebias* (Costa 2004) contains 4–8 named species (depending on the degree of taxonomic splitting) that constitute a distinct clade of rivuline killifishes, Rivulidae. Recent molecular-genetic data have illuminated the phylogenetic relationships of these species and also clarified their modes of reproduction (Tatarenkov et al. 2009). In particular, microsatellite analyses documented a high selfing rate (97%) in a nominal species (*K. ocellatus*), which, based on mtDNA sequences and other evidence, is the sister taxon (closest living relative) to *K. marmoratus* (fig. 4.23). In contrast, the microsatellite data uncovered no evidence of self-fertilization in *K. caudomarginatus* (an androdioecious species closely related to the *marmoratus-ocellatus* clade), and they confirmed (as expected) that outcrossing is the norm in *K. brasiliensis* (a phylogenetic outlier species with separate sexes). Overall, these genetic findings imply that the evolutionary origin of the capacity for self-fertilization in the Rivulidae predated the origin of the *marmoratus/ocellatus* clade, and the genetic data also thereby indicate that the selfing capacity has persisted in these fishes (as part of a mixed-mating system) for at least several hundred thousand years (based on molecular-clock considerations for mtDNA).

History of Genetic Research on the Mangrove Rivulus

The remarkable dual-sex reproductive mode of *K. marmoratus* was discovered a half century ago by Robert Harrington (Harrington 1961) and has since been the subject of many detailed genetic and evolutionary analyses (reviewed in Avise 2008). Early on, Harrington and colleagues found that they could successfully graft fins and organs between a hermaphroditic individual and its offspring, or between progeny within a sibship, thus indicating that the fish were genetically identical and probably homozygous at histocompatibility loci (Kallman and Harrington 1964; Harrington and Kallman 1968). By contrast, artificial grafts between some of the inbred lines were rejected acutely by the recipient fish, thus implying that particular selfing strains were genetically different. Harrington and Kallman used the word *clone* to refer to each highly inbred strain of *K. marmoratus*, a practice that continues today. This usage is not without pitfalls, however, because "clone-mates" arising via continued selfing could be misconstrued to have arisen via non-meiotic processes (as in parthenogenesis) and also because the genetic delimitation of a "clone" can be ambiguous when (as often happens) refined molecular assays uncover cryptic genetic variation within a previously suspected clonal entity.

Each decade since the 1960s has witnessed the introduction of new laboratory methods to assay DNA and proteins in ever-closer detail at the molecular level (Avise 2004), and many of these approaches have been applied over the years to the Mangrove Rivulus. These include protein-electrophoretic assays (Massaro et al. 1975; Vrijenhoek 1985), multi-locus DNA fingerprinting (Turner et al. 1990, 1992b; Laughlin et al. 1995), mtDNA restriction site analyses (Weibel et al. 1999), molecular surveys of loci in the major

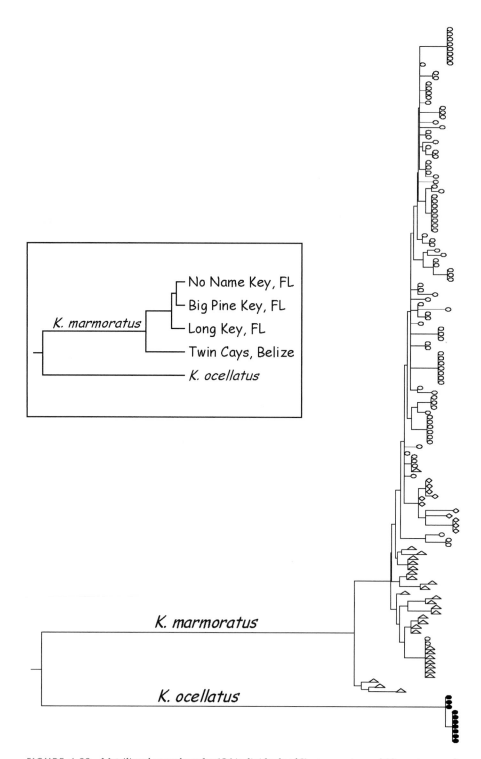

FIGURE 4.23 Matrilineal genealogy for 136 individuals of *K. marmoratus* and 10 specimens of *K. ocellatus* based on mtDNA sequence data (after Tatarenkov et al. 2009). Open ovals, triangles, and rhombi designate individuals from different populations of *K. marmoratus*. *Inset*: Phenogram for various populations of these two species based on genetic data from 31 microsatellite loci.

histocompatibility complex (MHC) (Sato et al. 2002), appraisals of multi-locus microsatellite loci (Mackiewicz et al. 2006a,b,c), and, most recently, mtDNA sequencing (Murphy et al. 1999; Lee et al. 2001; Tatarenkov and Avise 2007; Tatarenkov et al. 2007, 2009). Various of these molecular approaches generally have confirmed that natural populations of *K. marmoratus* often consist of strains that are so inbred as to be, in effect, clonal. However, some of the assays have revealed far more genetic variation in *K. marmoratus* than formerly had been appreciated. Furthermore, by revealing the sources of this genetic variation (including outcrossing as part of a mixed-mating system), these latest molecular techniques also have identified previously unsuspected genetic and reproductive phenomena in the species.

For example, Floridian populations of *K. marmoratus* often show near-zero heterozygosity (within-individual genetic variation) but extensive between-individual genetic variation (clonal diversity) (table 4.3). The low heterozygosity clearly is due to intense inbreeding that accompanies multi-generation selfing. Previously, the clonal diversity at particular geographic locations had been attributed to *de novo* mutations and interlocality gene flow (Turner et al. 1990; Laughlin et al. 1995), but the newer microsatellite data have made it abundantly clear that occasional outcrossing is also a major contributing factor (Mackiewicz et al. 2006a,b,c; Tatarenkov et al. 2007). The joint pattern of near-zero heterozygosity yet extensive "clonal" genetic diversity within local populations of *K. marmoratus,* due to the species' mixed-mating system, has given rise to the "fireworks model" (box 4.4) for this peculiar population-genetic pattern.

Adaptive Significance of Selfing

At least in theory, one advantage of selfing is intrinsic to the process: a selfer transmits two sets of genes to each offspring, whereas an outcrosser transmits only one set (Nagylaki 1976; Lloyd 1979b). Other potential benefits from selfing have more to do with ecological or behavioral considerations. Thus, as described in chapters 2 and 3 for plants and invertebrate animals, respectively, two primary hypotheses have been advanced for how high selfing rates might pay regular fitness dividends as a key component of a mixed-mating system. First, the intense inbreeding associated with regular selfing allows at least the potential for the perpetuation and proliferation of co-adapted multi-locus genotypes that might be highly adapted to particular ecological conditions. Under this view, consistent selfing might often be advantageous in the ecological short term because it can yield progeny with identical (effectively clonal) copies of potentially coadapted multi-locus genotypes that nature already has field-tested for genetic fitness (in parental lineages) in particular habitats. However, outcrossing remains an important

TABLE 4.3 Example of clonal diversity in a Floridian population of *K. marmoratus*. Each column is a microsatellite locus, and each row is a different fish specimen. The body of the table shows each individual's diploid genotype at each locus (numbers refer to the sizes of different alleles). Note that each individual has a unique multi-locus genotype that nonetheless is highly homozygous.

				Locus				
Specimen	1	2	3	4	5	6	7	8
a	154/154	310/310	198/198	224/224	219/219	174/174	284/284	278/278
b	156/156	310/310	198/198	224/224	219/219	174/174	284/284	278/278
c	154/154	312/312	198/198	224/224	219/219	174/174	284/284	278/278
d	154/154	310/310	200/200	224/224	219/219	174/174	284/284	278/278
e	162/162	294/294	198/198	224/224	204/204	178/178	284/284	286/286
f	162/162	294/294	198/198	224/224	204/204	178/178	284/284	286/286
g	166/166	298/298	198/198	224/224	228/228	166/166	284/284	278/278
h	134/134	302/302	198/198	224/224	213/213	166/166	284/284	274/274
i	158/158	302/302	198/198	224/224	216/216	174/174	284/284	282/282

BOX 4.4   The Fireworks Model

The documentation of at least occasional outcrossing against a backdrop of predominant selfing gives rise to what has been named the "fireworks model" (Avise 2008) for the population genetic architecture of *K. marmoratus* at particular sites such as in Florida. In this metaphor, a black nighttime sky represents the near-complete absence of within-individual heterozygosity in an inbred (highly selfed) population, and each exploding firework represents a single outcross event between distinct homozygous clones. At the core of each explosion is a bright spot of light that represents high heterozygosity in the outcross progeny. Streamers of light, brilliant at first but then quickly fading, burst out from this core as the highly heterozygous offspring begin to reproduce, often by a resumption of selfing. The many streamers of light that head in different directions represent the many different recombinant genotypes that inevitably arise during this reproductive process, but the streamers quickly fade back into darkness as intra-strain heterozygosity is lost in each successive generation of selfing. Then another explosion occurs, perhaps in a different part of the nighttime sky, as another outcross event between distinct clonal lineages releases another brilliant but temporary burst of genetic variation available for recombination.

component of the mixed-mating system as well, especially when habitats change over time or show spatial heterogeneity, because outcrossing produces genetically diverse progeny, some of which may be well suited to the new environmental regimes. Second, selfing provides additional fertilization insurance that simply is unavailable to gonochorists or even to outcrossing simultaneous hermaphrodites. Against these potential advantages of selfing stands one huge potential disadvantage: the inbreeding depression that generally is expected within lineages that constitutively or regularly self-fertilize. On balance, however, a mixed-mating system might convert the inbreeding dilemma of constitutive selfing into a best-of-both-worlds adaptive strategy that combines many of the advantages of sexual and clonal reproduction (Allard 1975).

For *K. marmoratus*, the first of these two potential advantages of selfing seems unlikely to be applicable, because the genetic data offer no empirical support for the idea that particular multi-locus genotypes are either common (e.g., table 4.3) or associated with particular ecological conditions. This leaves fertilization insurance as the most compelling hypothesis for any selective advantage that selfing routinely affords the Mangrove Rivulus.

Ghiselin (1969:189) encapsulated this latter adaptive consideration for self-fertilizing hermaphrodites when he wrote, "The necessity to reproduce at all costs should favor the development of selfing wherever the environment is such that the transfer of gametes between individuals is hindered." In effect, Gheselin's statement is a partial reinvocation of Baker's rule (chapter 2), which states that natural selection should favor self-fertilization capabilities especially in dispersive or weedy species, because even a single selfing individual can be a successful colonist.

The behavior and natural history of the Mangrove Rivulus can be interpreted as consistent with Baker's rule in several regards: the species has a large geographic range, extending from Brazil to Florida and including many Caribbean islands; individuals tend to occupy mangrove litter and termite cavities in rotting logs and thus may be behaviorally and ecologically predisposed to occasional long-distance passive dispersal via floating forest litter (i.e., as flotsam and jetsam following storms); adults can survive out of water for up to 10 weeks; fertilized ova are well suited for dispersal because they too can survive out of water for long periods; and many rivulus individuals lead rather isolated, independent lives.

All of these biological attributes of *K. marmoratus* imply that selfing should often be advantageous for individuals in this species because of the fertilization assurance it confers. Thus, overall, the mixed-mating system of the Mangrove Rivulus probably combines some of outcrossing's general advantages (maintenance of recombinant genetic variation and adaptability) with one of selfing's most obvious fitness benefits (fertilization insurance).

SUMMARY

1. Hermaphroditism is displayed in several hundred species of bony fishes (teleosts) distributed across about 30 taxonomic families in a dozen orders. Most of these dual-sex fishes are sequential hermaphrodites that change sex sometime in life after an initial stage either as a male (protandry) or as a female (protogyny). All of the higher taxa in which hermaphroditism is represented consist primarily of gonochoristic species, thus strongly implying that hermaphroditism in fishes is a polyphyletic condition normally derived from an ancestral condition of separate sexes. Unlike the case in plants, the transitional evolutionary stages between gonochorism and hermaphroditism remain poorly known in fishes, in large part because intermediate phases are rare in extant fish taxa.

2. The fact that many fish lineages display hermaphroditism whereas other vertebrate lineages do not is testimony to the sexual flexibility of fishes vis-à-vis mammals, birds, and many reptiles and amphibians. Such flexibility has two interrelated aspects: evolutionary and ontological. The evolutionary aspect of flexibility is registered by the polyphyletic nature of hermaphroditism in fishes, and also by the relatively rapid evolutionary interconversions between sexual modes (such as alternative sex-determining mechanisms) and between different expressions of hermaphroditism (such as protogyny versus protandry). Sequential hermaphroditism itself is prima facie evidence for the ontogenetic aspect of sexual flexibility in fishes.

3. Sequential hermaphroditism is displayed by a wide variety of marine fishes, most notably by many species that inhabit coral reefs, but also in miscellaneous other groups including some bathypelagic (deep-sea) taxa. Collectively, sequential hermaphrodites display a fascinating diversity of life histories and natural histories. For example, many groupers and wrasses are protogynous, with many of the latter having haremic lifestyles, male and/or female dominance hierarchies, and socially mediated sex changes. Other species such as some damselfishes are protandrous, and many marine gobies show serial sex changes in both directions during a typical lifetime.

4. In prior decades, researchers sometimes invoked "group-selection" arguments to account for various expressions of sequential hermaphroditism in fishes. However, more modern analyses have focused on ecological and demographic conditions that might behoove individuals to change sex at particular times in life so as to enhance personal prospects for current or future reproductive success. Especially useful has been the "size-advantage model" which predicts the optimal time of sex change by considering how the age-specific or the size-specific fecundities of males and females vary during

ontogeny as a function of each individual's ecological and demographic circumstances. The size-advantage model has proved to be powerful for addressing why some sequentially hermaphroditic fish species are pro-togynous, others are protandrous, and others are serial bi-directional sex changers.

5. Conventional wisdom is that fitness trade-offs related to ecological, be-havioral, and life-history traits influence whether protandrous or protogy-nous hermaphroditism evolves in a given fish species. One general consid-eration is that sperm are individually small and inexpensive to produce compared to eggs, such that body size typically limits female fecundity far more than it does male fertility. Another general consideration is that most fish have more-or-less indeterminate growth, meaning that body size can increase throughout life. Thus, if all else were equal, it might behoove an individual to produce sperm when young and small, but perhaps to switch to egg production after a larger body size has been attained. Protandry is often interpreted as an outcome of this kind of selection pressure. On the other hand, males in many fish species defend scarce territories or otherwise compete intensely for female access, so only larger specimens might expect high reproductive success as males. In such circumstances, protogyny might tend to be the selectively favored evolutionary outcome, and, indeed, pro-togyny is the most common form of sequential hermaphroditism in fish.

6. Synchronous hermaphroditism—in which an individual simultaneously produces eggs and sperm—is relatively uncommon in fishes, being known only in about 40 species in half-a-dozen taxonomic families. Most of these simultaneous hermaphrodites are outcrossers rather than selfers. During a typical spawning episode in some such species, each dual-sex individual in-tersperses male and female roles in close succession, alternately releasing eggs and sperm in a "tit-for-tat" mating game with its partner.

7. Members of only one fish clade (containing the Mangrove Rivulus and its sister taxon, both in the genus *Kryptolebias*) are known to self-fertilize rou-tinely. Most individuals have an ovotestis that produces both eggs and sperm, which unite inside the hermaphrodite's body. Selfing in *Kryplolebias* is actually part of a mixed-mating system that also includes outcrossing (at various frequencies, depending on the population) probably mediated by occasional males that also exist in these androdioecious taxa. The adaptive significance of selfing in *Kryptolebias* appears to be related to fertiliza-tion insurance in these relatively solitary yet highly dispersive fishes, thus making these animals compliant with Baker's rule as originally developed for self-fertilizing plants and invertebrate animals with high colonization potential.

GLOSSARY

adaptation Any feature (e.g., morphological, physiological, behavioral) that helps an organism to survive and reproduce in a particular environment.

aging *See* senescence.

allele Any of the possible forms (or classes of forms) of a specified gene. A diploid individual carries two alleles at each autosomal gene, and these can either be identical in state (in which case the individual is homozygous) or different in state (heterozygous). At each autosomal gene, a population of N diploid individuals harbors $2N$ alleles, various of which may differ in details of nucleotide sequence.

allohormone Any substance, transferred directly into the body of a conspecific, that induces a direct physiological response.

allopatric Inhabiting different geographic areas.

allozyme A genetic form of a protein revealed by protein electrophoresis.

androdioecy A condition in which hermaphrodites and males coexist within a species.

androgyny *See* hermaphroditism.

andromonoecy A condition in which a plant bears both male and bisexual flowers.

angiosperm A flowering plant with fruit-encased seeds.

anisogamy   The bimodal distribution of gamete size (smaller in males, larger in females) that characterizes nearly all sexually reproducing organisms.

anther   The terminal portion of a stamen.

apomixis   Asexual reproduction without meiosis or fertilization.

asexual reproduction   Any form of reproduction that does not involve the fusion of sex cells (gametes).

autogamy   Fertilization entailing the same flower on a plant, or the same polyp on a coral.

autosome   A chromosome in the nucleus other than a sex chromosome; in diploid organisms, autosomes are present in homologous pairs.

Baker's rule   The tendency for hermaphroditic individuals in weedy or dispersive species to be self-fertilizers due to the fertilization insurance that selfing provides.

Bateman gradient   A statistical regression describing the relationship between mate numbers and progeny production.

bathypelagic   Pertaining to the deep sea.

biodiversity   Life's genetic heterogeneity, at any or all levels of biological organization.

bisexual   A population composed of separate male and female individuals; or, an individual or a component thereof (such as a flower) with both male and female parts.

bottleneck   A severe but temporary reduction in population size.

carpel   The female reproductive structure in a flowering plant, typically consisting of a basal ovary, elongate style, and terminal stigma.

cell   A small, membrane-bound unit of life that is usually capable of self-reproduction.

chloroplast   An organelle in the cellular cytoplasm that is involved in photosynthesis in many plants.

chloroplast DNA (cp DNA)   A small, typically circular genome housed within the chloroplast.

chasmogamy   referring to flowers that are open and adapted to fertilization by outcrossing. *See also* cleistogamy.

chromosome   A threadlike structure within a cell that carries genes.

clade   A monophyletic group of organisms.

classification (biological)   A process of establishing, defining, and ranking biological taxa within hierarchical groups; or, the outcome itself of this process.

cleistogamous   Pertaining to cleistogamy.

cleistogamy ("closed marriage")   referring to flowers in which self-fertilization occurs within a closed bud. *See also* chasmogamy.

clonal   Pertaining to a clone.

clone   (*n*) a biological entity (e.g., gene, cell, or multicellular organism) that is genetically identical to another; or, all genetically identical entities that have descended asexually from a given ancestral entity; (*v*), to produce such genetically identical entities or lineages.

clonemates   Two or more organisms that are genetically identical.

clutch   The eggs or offspring within a brood.

coadaptation   The joint adaptation to one another of two or more biological entities.

coevolution   The interdependent evolution of two or more interacting species or genomes.

congeneric   Belonging to the same taxonomic genus.

conifer   A type of gymnosperm plant that typically bears cones.

conspecific   Belonging to the same taxonomic species.

constitutive   Of consistent and essential occurrence. *See also* facultative.

cosexual   Hermaphroditic.

cuckoldry   Stolen copulations often resulting in extra-pair paternity.

cytonuclear analysis   A genetic appraisal based on information jointly from an organism's cytoplasmic genome (typically mtDNA) and one or more loci in the nuclear genome.

cytoplasm   The portion of a cell outside the nucleus.

cytoplasmic male sterility   Infertility of males due to genetic factors housed in the cellular cytoplasm.

deme   A local interbreeding population.

deoxyribonucleic acid (DNA)   The genetic material of most life forms; a double-stranded molecule composed of strings of nucleotides.

dichogamy   A condition in which carpels and stamens mature at different times during a plant's development.

dioecy   A condition in which males and females are separate individuals (often used in the botanical literature). *See also* gonochorism.

diploid   A usual somatic cell condition wherein two copies of each chromosome are present.

distyly   A common type of heterostyly in which different conspecific plants have two distinct flower morphs.

ecology   The study of the interrelationships among living organisms and their environments.

ecotone   A transitional zone between adjacent ecological communities.

ectoparasite   A parasite that remains on the outside of a host's body.

effective population size   The number of individuals in an idealized population displaying the same genetic properties as those observed in the actual population under consideration.

egg   A female gamete; oocyte.

electrophoresis   The movement of charged proteins or nucleic acids through a supporting gel under the influence of an electric current.

embryo   An organism in the early stages of development.

embryogenesis   The development of an embryo.

endemic   Native to, and restricted to, a particular geographic area.

endogenous   Produced or naturally occurring within the body.

endoparasite   A parasite that resides inside a host's body for at least part of its life cycle.

endosymbiosis   An intimate symbiotic relationship that begins when one organism takes up perhaps permanent and mutually beneficial residence within another.

enzyme   A catalyst (normally a protein) of a specific chemical reaction.

epigamic selection   A form of sexual selection that operates via mating preferences of females for particular males (or sometimes vice versa).

epigenetic   Any mechanism during ontogeny that causes phenotypic variation without altering the nucleotide sequences of the genes.

ethology   The study of animal behavior.

eukaryote   Any organism in which chromosomes are housed in a membrane-bound nucleus.

evolution   Change across time in the genetic composition of a population or species.

evolutionary stable strategy   The value at which a trait in a population is immune to permanent invasion by an alternative trait value.

exogenous   Produced or naturally occurring outside the body.

expression (of a gene)   Activation of a gene to begin the process (RNA formation) that later may eventuate in production of a protein.

extant   Alive today.

extinction   The permanent disappearance of a population or species.

facultative   Optional; occurring only part of the time. *See also* constitutive.

fecundity   The number of gametes an individual produces; fertility.

female   The sex that produces relatively large gametes.

fertility   *See* fecundity.

fertilization   The union of two gametes to produce a zygote; syngamy.

fitness (Darwinian)   The contribution of an individual or a genotype to the next generation relative to the contributions of other individuals or genotypes. *See also* inclusive fitness.

fixation   A situation in which an allele has reached 100% frequency in a population.

frequency-dependent selection   A form of natural selection that varies as a function of the population frequencies of alternative genotypes.

gamete   A mature reproductive sex cell (egg or sperm).

game theory   A branch of theoretical population biology in which researchers analyze how animals optimally act when an individual's success depends on others' decisions as well as its own.

gametogenesis   The process by which sex cells are produced.

gametophyte   The haploid sexual phase of a plant.

geitonogamy   Fertilization entailing different flowers on the same plant.

gene   The basic unit of heredity; usually taken to imply a sequence of nucleotides specifying production of a polypeptide or other functional product, but also can be applied to stretches of DNA with unknown or unspecified function.

genealogy   A record of descent from ancestors through a pedigree.

gene flow   The geographic movement of genes, normally among populations within a species.

gene pool   The sum total of all hereditary material in a population or species.

genetic   Pertaining to the study of heredity.

genetic drift   Change in allele frequency in a finite population by chance sampling of gametes between generations.

genetic load   The burden to a population of deleterious genes.

genetic marker   Any of the natural nucleic-acid or protein tags that exist in all forms of life.

genome   The complete genetic constitution of an organism; also can refer to a particular composite discrete piece of genetic material such as mtDNA or cpDNA.

genotype   The genetic constitution of an individual with reference to a gene or set of genes.

germline   The lineage of cells leading to an individual's gametes.

gonochorism   A sexual system in which each individual is either a male or a female (often used in the zoological literature). *See also* dioecy.

group selection   Natural selection acting upon groups of individuals via differences in the traits that those groups possess.

gymnosperm   A plant with naked seeds not enclosed in a fruit.

gynodioecy   A condition in which hermaphrodites and females both occur within a species.

gynomonoecy   A condition in which a plant bears both female and bisexual flowers.

haploid   The usual condition of a gametic cell in which one copy of each chromosome is present.

harem   A group of females that mate with a particular male.

haremic   Pertaining to a harem.

herbaceous   Pertaining to non-woody plants whose stems are not thickened and lignified.

heredity   The phenomenon of familial transmission of genetic material from one generation to the next.

herkogamy (hercogamy)   pertaining to a flower having stamens and stigma positioned in such a way as to prevent self-fertilization.

hermaphroditism   A condition in which an individual produces both male and female gametes, either sequentially or simultaneously.

heterogametic sex   The gender that produces gametes each containing one of two different types of sex chromosomes.

heterosis   Higher genetic fitness of heterozygotes than homozygotes.

heterospecific   Pertaining to another species.

heterostyly   The physical separation of carpels and stamens within a perfect flower.

heterozygosity   The percentage of heterozygotes, or the percentage of loci in heterozygous state, in a population.

heterozygote   A diploid organism possessing two different alleles at a specified gene.

histocompatibility   Pertaining to genes that influence the acceptance or rejection of cells or tissues in grafts.

homogametic sex   The gender that produces gametes all containing the same type of sex chromosome.

homology   Similarity of traits (morphological, molecular, etc.) due to inheritance from a shared ancestor.

homospecific   Pertaining to the same species.

homozygote   A diploid organism possessing two identical alleles at a specified gene.

hormone   A chemical substance produced by the body that produces a specific physiological response.

hybridization   The successful mating of individuals belonging to genetically different populations or species.

inbreeding   Mating and reproduction between kin.

inbreeding depression   A loss in genetic fitness due to inbreeding.

incestuous   Pertaining to matings between close kin.

inclusive fitness   An individual's own genetic fitness as well as his or her effects on the genetic fitness of close relatives.

inflorescense   The arrangement of flowers on a flowering shoot.

intersexual   Showing phenotypic features intermediate to male and female.

introgression   The movement of genes between species via hybridization.

invertebrate   An animal that does not possess a backbone.

isogamy   A condition entailing the fusion of gametes of similar size. *See also* anisogamy.

kin selection   A form of natural selection due to individuals favoring the survival and reproduction of genetic relatives.

life cycle   The sequence of events for an individual, from its origin as a zygote to its death; one generation.

linked genes   Loci carried on the same chromosome.

locus (pl. loci)   A gene, a location on a chromosome.

male   The sex that produces relatively small gametes.

mating system   The general pattern by which males and females mate (or male and female gametes unite) within a population or species. *See also* monogamy, polygamy, polyandry, polygyny, polygynandry, promiscuity, selfing, outcrossing.

matriline   A genetic transmission pathway strictly through females.

meiosis   The cellular process whereby a diploid cell divides to form haploid gametes.

meiotic   Pertaining to meiosis.

meiotic drive   Any tendency for a particular allele to preferentially be distributed to gametes during meiosis.

meristem   the undifferentiated, mitotically active tissues of plants.

metabolism   The sum of all physical and chemical processes by which living matter is produced and maintained, and by which cellular energy is made available to an organism.

microbe   A very small organism visible only under a microscope.

microsatellite   A locus containing tandem repeats of short nucleotide sequences.

mitochondrial DNA (mtDNA)   A small, typically circular genome housed within the mitochondrion.

mitochondrion (pl. mitochondria)   An organelle in the cytoplasm of animal and plant cells that is the site of some key metabolic pathways involved in cellular energy production.

mitosis   A process of cell division that produces daughter cells with the same chromosomal constitution as the parent cell.

mitotic   Pertaining to mitosis.

mixed-mating   A mating system that involves both selfing and outcrossing.

molecular clock   An evolutionary timepiece based on the evidence that genes or proteins tend to accumulate mutational differences at roughly constant rates in particular lineages.

molecular marker   *See* genetic marker.

monandric   Pertaining to a form of protandry in which all males derive from sex-changed females.

monoecy   A situation in which an individual plant has male and female reproductive organs and produces both pollen and ova.

monogamy   A mating system in which each male and each female has only one mate.

monophyletic   Of single evolutionary origin.

morphology   The visible structures of organisms.

multicellular   Composed of two or more cells.

mutation   A change in the genetic constitution of an organism.

natural history   The study of nature including organisms and natural phenomena.

natural selection   The differential contribution by individuals of different genotypes to the next generation.

nepotism   Favortism directed toward kin.

niche   The ecological "role" of a species in a natural community; an organism's way of making a living.

nucleic acid   *See* deoxyribonucleic acid and ribonucleic acid.

nucleotide   A unit of DNA or RNA consisting of a nitrogenous base, a pentose sugar, and a phosphate group.

nucleus (pl. nuclei)   A portion of a cell bounded by a membrane and containing chromosomes.

ontogeny   The course of development and growth of an individual to maturity.

oocyte (unfertilized)   A female gamete, also known as an egg cell, or ovum.

oogenesis   The production of oocytes.

operational sex ratio   The number of males versus females (or their gametes) effectively available for reproduction during the time period under consideration.

organ   A part of an animal, such as the heart, that forms a structural and functional unit.

organelle   A complex, recognizable structure in the cell cytoplasm (such as a mitochondrion or chloroplast).

outbreeding depression   A loss in genetic fitness in hybrids due to negative interactions between genes that had differentiated in allopatric populations or different species.

outcrossing   Mating with another, typically unrelated, individual.

ovary   An egg-producing organ.

overdominance   A situation in which heterozygotes have higher genetic fitness than homozygotes.

oviparous   Egg-laying.

ovotestis   An organ that produces both eggs and sperm.

ovule   The structure in seed plants that develops into a seed after fertilization of the egg within it.

ovum   Egg.

parthenogen   An individual or strain that reproduces by parthenogenesis.

parthenogenesis   The development of an individual from an egg without fertilization.

pelagic   Pertaining to the open ocean.

pedigree   A diagram displaying mating partners and their offspring across generations.

perfect flower   A flower with both male and female functional parts.

phenotype   Any morphological, physiological, behavioral, or other such characteristics of an organism.

phenotypic plasticity   Variation among phenotypes not due directly to genetic differences.

phylogeny   Evolutionary relationships (historical descent) of a group of organisms or species.

phylogenetic   Pertaining to phylogeny.

phylogenetic character mapping (PCM)   A scientific exercise in which alternative traits are plotted and ancestral states are inferred in a phylogenetic framework.

pistil   The ovary of a flower plus its style and stigma.

pistillate   Having a pistil but no functional stamens; female.

pleiotropy   A phenomenon in which a single gene can contribute to more than one phenotype.

pollen   A male gamete in plants.

pollen discounting   A reduction in the amount of pollen available for export and cross-fertilization.

pollination   The transference of pollen to receptive female parts of a plant.

polyandry   A mating system in which particular females may have multiple mates but each male typically has only one mate.

polygamy   A mating system in which at least some individuals have multiple mates. *See also* polyandry, polygyny, polygynandry, promiscuity.

polygyny   A mating system in which particular males may have multiple mates but each female typically has only one mate.

polygynandry   A mating system in which both males and females may have several mates each.

polymorphism   The presence of two or more genetically distinct forms (traits or genotypes) in a population.

polyphyletic   A group of organisms perhaps classified together but tracing to different ancestors.

polyploid   A condition in which more than two sets of chromosomes are present within a cell.

population   All individuals of a species normally inhabiting a defined area.

prokaryote   Any microorganism that lacks a chromosome-containing, membrane-bound nucleus.

promiscuity   An extreme form of polygynandry in which each male and female has many mating partners.

protandry   A type of hermaphroditism in which an individual is first male and then later in life switches to female.

protein   A macromolecule composed of one or more polypeptide chains.

protogyny   A type of hermaphroditism in which an individual is first female and then later in life switches to male.

pseudohermaphroditism   The false appearance, due to environmentally induced phenotypes, of hermaphroditism in an invertebrate species that otherwise is genetically gonochoristic.

recombination (genetic)   The formation of new combinations of genes, as for example occurs naturally via meiosis and fertilization.

regulatory gene   A segment of DNA that exerts operational control over the expression of other genes.

ribonucleic acid (RNA)   The genetic material of many viruses, similar in structure to DNA. Also, any of a class of molecules that normally arise in cells from the transcription of DNA.

secondary sexual traits   Phenotypic characters other than the primary reproductive organs that are confined to or are elaborated in only one sex and that evolved via sexual selection.

self-fertilization (selfing)   The union of male and female gametes from the same hermaphroditic individual.

self-incompatibility   Any incapacity of a hermaphrodite to fertilize itself.

senescence   A persistent decline with age in the survival probability or reproductive output of an individual due to interior physiological deterioration.

sepal   Each element in the outermost protective whorl of a typical flower.

sex allocation   The relative parental investment in male-versus-female reproductive functions.

sex chromosome   A chromosome in the cell nucleus involved in distinguishing the two genders.

sex ratio   The relative number of males versus females in a population.

sexual dimorphism   Consistent differences in secondary sexual traits between males and females.

sexual reproduction   Organismal procreation via the generation and fusion of gametes.

sexual selection   Selection pressures arising from intraspecific competition for mates.

sexual selection gradient   *See* Bateman gradient.

soma   *See* somatic.

somatic   Of or pertaining to any cell (or body part) in a multicellular organism other than those destined to become gametes.

species (biological)   Groups of actually or potentially interbreeding individuals that are reproductively isolated from other such groups.

sperm   A male gamete in animals.

spermatid   One of the four haploid cells produced during meiosis in males.

sperm competition   Competition among sperm for fertilization success.

spermatheca   A female storage organ for sperm.

spermatogenesis   The production of sperm.

spermatophore   A packet of sperm.

stamen   Each of the male reproductive organs in a plant.

staminate   Having stamens but not carpels; male.

stigma   The receptive apex of a flower's female parts; the site that receives pollen.

style   The elongate portion of a flower's female parts, between the ovary and stigma.

subdioecy   *See* trioecy.

sympatric   Inhabiting the same geographic area.

syngamy   The genetic union of a male gamete and a female gamete.

systematics   The comparative study and classification of organisms, particular with regard to their phylogenetic relationships.

taxon (pl. taxa)   A biotic lineage or entity deemed sufficiently distinct from other such lineages as to be worthy of a formal taxonomic name.

taxonomy   The practice of naming and classifying organisms.

testis   A sperm-producing organ.

tetraploid   Possessing four complete sets of chromosomes.

tissue   A population of cells of the same type performing the same function.

transcription   The cellular process by which an RNA molecule is formed from a DNA template.

translation   The cellular process by which a polypeptide chain is formed from an RNA template.

trioecy   A condition in which hermaphrodites, males, and females all occur within a species.

triploid   Possessing three complete sets of chromosomes.

tristyly   A type of heterostyly in which three distinct morphs of perfect flowers coexist within a species.

variance   statistical variation; the mean squared deviation from the mean.

vertebrate   An animal that possesses a backbone.

W-chromosome   In birds, the sex chromosome normally present in females only.

X-chromosome   The sex chromosome normally present as two copies in female mammals (the homogametic sex), but as only one copy in males (the heterogametic sex).

Y-chromosome   In mammals, the sex chromosome normally present in males only.

Z-chromosome   The sex chromosome normally present as two copies in male birds (the homogametic sex), but as only one copy in females (the heterogametic sex).

zygote   Fertilized egg; the diploid cell arising from the union of male and female haploid gametes.

REFERENCES CITED

Abe M, and Fukuhara H. 1996. Protogynous hermaphroditism in the brackish and freshwater isopod, *Gnorimosphaeroma naktongense* (Crustacea: Isopoda, Sphaeromatidae). *Zool. Sci.* 13: 325–29.

Adamo SA, and Chase R. 1988. Courtship and copulation in the terrestrial snail *Helix aspersa*. *Can. J. Zool.* 66: 1446–53.

Adamo SA, and Chase R. 1996. Dart shooting in helicid snails: An "honest" signal or an instrument of manipulation? *J. Theoret. Biol.* 180: 77–80.

Aide TM. 1986. The influence of wind and animal pollination on variation in outcrossing rates. *Evolution* 40: 434–35.

Akimoto J, Fukuhara R, and Kikuzawa K. 1999. Sex ratios and genetic variation in a functionally androdioecious species, *Schizopepon bryoniaefolius* (Cucurbitaceae). *Amer. J. Bot.* 86: 880–86.

Allard RW. 1975. The mating system and microevolution. *Genetics* 79: 115–26.

Allard RW, Babbel GR, Clegg MT, and Kahler AL. 1972. Evidence for coadaptation in *Avena barbata*. *Proc. Natl. Acad. Sci. USA* 69: 3043–48.

Allen GR. 1975. *Anemonefishes*. 2d ed. Melle, Germany: Mergus.

Allen SK, Hidu H, and Stanley JG. 1986. Abnormal gametogenesis and sex ratio in triploid soft-shell clams (*Mya arenaria*). *Biol. Bull.* 170: 198–220.

Allsop DJ, and West SA. 2003. Constant relative age and size at sex change for sequentially hermaphroditic fish. *J. Evol. Biol.* 16: 921–29.

Allsop DJ, and West SA. 2004. Sex-ratio evolution in sex changing animals. *Evolution* 58: 1019–27.

Anderson GJ, Bernardello G, Stuessy TF, and Crawford DJ. 2001. Breeding system and pollination of selected plants endemic to Juan Fernandez Islands. *Amer. J. Bot.* 88: 220–33.

Anderson GJ, and Symon DE. 1989. Functional dioecy and andromonoecy in *Solanum*. *Evolution* 43: 204–219.

Andersson M. 1994. *Sexual Selection*. Princeton: Princeton UP.

Andersson M, and Isawa Y. 1996. Sexual selection. *Trends Ecol. Evol.* 11: 53–58.

Angeloni L. 2003. Sexual selection in a simultaneous hermaphrodite with hypodermic insemination: Body size, allocation to sexual roles and paternity. *Anim. Behav.* 66: 417–26.

Angeloni L, Bradbury JW, and Charnov EL. 2002. Body size and sex allocation in simultaneously hermaphroditic animals. *Behav. Ecol.* 13: 419–26.

Anthes N, and Michiels NK. 2005. Do "sperm trading" simultaneous hermaphrodites always trade sperm? *Behav. Ecol.* 16: 188–95.

Anthes N, Putz A, and Michiels NK. 2005. Gender trading in a hermaphrodite. *Current Biol.* 15: R792–93.

Anthes N, Putz A, and Michiels NK. 2006a. Sex role preferences, gender conflict and sperm trading in simultaneous hermaphrodites: A new framework. *Anim. Behav.* 72: 1–12.

Anthes N, Putz A, and Michiels NK. 2006b. Hermaphrodite sex role preferences: The role of partner body size, mating history and female fitness in the sea slug *Chelidonura sandrana*. *Behav. Ecol. Sociobiol.* 60: 359–67.

Arbuthnot J. 1710. An argument for Divine Providence, taken from the constant Regularity observ'd in the Births of both Sexes. *Phil. Trans. Royal Soc. London* 27: 186–90.

Arkhipchuk VV. 1995. Role of chromosomal and genome mutations in the evolution of bony fishes. *Hydrobiologia* 31: 55–65.

Armstrong CN., and Marshall AJ., eds. 1964. *Intersexuality in Vertebrates Including Man*. New York: Academic Press.

Arnold SJ, and Duvall D. 1994. Animal mating systems: A synthesis based on selection theory. *Amer. Natur.* 143: 317–48.

Arnqvist G, and Rowe L. 2005. *Sexual Conflict*. Princeton: Princeton UP.

Arthur DR. 1950. Abnormalities in the sexual apparatus of the common dogfish (*Scyliorhinus caniculus*). *Proc. Linn. Soc. Lond.* 162: 52–56.

Ashman T-L. 1994. Reproductive allocation in hermaphroditic and female plants of *Sidalcea oregana* ssp. *spicata* (Malvaceae) using four currencies. *Amer. J. Bot.* 81: 433–38.

Ashman T-L. 2002. The role of herbivores in the evolution of separate sexes from hermaphroditism. *Ecology* 83: 1175–84.

Ashman T-L. 2006. The evolution of separate sexes: A focus on the ecological context. In L. D. Harder and S. C. H. Barrett, eds., *Ecology and Evolution of Flowers*, 204–222. New York: Oxford UP.

Asikainen E, and Mutikainen P. 2003. Female frequency and relative fitness of females and hermaphrodites in gynodioecious *Geranium sylvaticum* (Geraniaceae). *Amer. J. Bot.* 90: 226–34.

Asoh K, and Yoshihawa T. 2003. Gonadal development and indication of functional protogyny in the Indian damselfish (*Dascyllus carneus*). *J. Zool.* 260: 23–39.

Atz JW. 1964. Intersexuality in fishes. In *Intersexuality in Vertebrates Including Man*, 145–232. New York: Academic Press.

Avise JC. 2002. *Genetics in the Wild*. Washington, D.C.: Smithsonian Institution Press.

Avise JC. 2004. *Molecular Markers, Natural History, and Evolution*. 2nd ed. Sunderland, Mass.: Sinauer.

Avise JC. 2006. *Evolutionary Pathways in Nature: A Phylogenetic Approach*. New York: Oxford UP.

Avise JC. 2008. *Clonality: The Genetics, Ecology, and Evolution of Sexual Abstinence in Vertebrate Animals*. New York: Oxford UP.

Avise JC. 2010. *Inside the Human Genome: A Case for Non-intelligent Design*. New York: Oxford UP.

Avise JC, and Mank JE. 2009. Evolutionary perspectives on hermaphroditism in fishes. *Sexual Develop.* 3: 152–63.

Avise JC, and 10 others. 2002. Genetic mating systems and reproductive natural histories of fishes: Lessons for ecology and evolution. *Annu. Rev. Genet.* 36: 19–45.

Avise JC, Power AJ, and Walker D. 2004. Genetic sex determination, gender identification and pseudohermaphroditism in the knobbed whelk, *Busycon carica* (Mollusca: Melongenidae). *Proc. Roy. Soc Lond B* 271: 641–46.

Axelrod R, and Hamilton WD. 1981. The evolution of cooperation. *Science* 211: 1390–96.

Baeza JA. 2006. Testing three models on the adaptive significance of protandric simultaneous hermaphroditism in a marine shrimp. *Evolution* 60: 1840–50.

Baeza JA. 2007. Sex allocation in a simultaneously hermaphroditic marine shrimp. *Evolution* 61: 2360–73.

Baeza JA, and Bauer RT. 2004. Experimental test of socially mediated sex change in a protandric simultaneous hermaphrodite, the marine shrimp *Lysmata wurdemanni* (Caridea: Hippolytidae). *Behav. Ecol. Sociobiol.* 55: 544–50.

Baeza JA, Reitz JM, and Collin R. 2007. Protandric simultaneous hermaphroditism and sex ratio in *Lysmata nayaritensis* Wicksten, 2000 (Decapoda: Caridea). *J. Natur. Hist.* 41: 2843–50.

Baker HG. 1948. Dimorphism and monomorphism in the Plumbagionaceae. I. Survey of the family. *Ann. Bot.* 12: 207–219.

Baker HG. 1953. Dimorphism and monomorphism in the Plumbagionaceae. III. Correlation of geographical distribution patterns with dimorphism and monomorphism in *Limonium*. *Ann. Bot.* 17: 615–27.

Baker HG. 1955. Self-compatibility and establishment after "long-distance" dispersal. *Evolution* 9: 347–49.

Baker HG. 1965. Characteristics and modes of origin of weeds. In H. G. Baker and G. L. Stebbins, eds., *Genetics of Colonizing Species*, 147–72. New York: Academic Press.

Baker HG, and Fox PA. 1984. Further thoughts on dioecism and islands. *Annals Missouri Bot. Gard.* 71: 244–53.

Baker RJ, and Bellis MA. 1995. *Human Sperm Competition*. London: Chapman & Hall.

Baldi C, Cho S, and Ellis RE. 2009. Mutations in two independent pathways are sufficient to create hermaphroditic nematodes. *Science* 326: 1002–1005.

Ballou JD. 1997. Effects of ancestral inbreeding on genetic load in mammalian populations. *J. Hered.* 88: 169–78.

Baras E, Jacobs B, and Melard C. 2001. Effects of water temperature on survival, growth and phenotypic sex of mixed (XX-XY) progenies of Nile tilapia *Oreochromis niloticus*. *Aquaculture* 192: 187–99.

Barlow GW. 1975. On the sociobiology of some hermaphroditic serranid fishes, the hamlets, in Puerto Rico. *Marine Biol.* 33: 295–300.

Baroiller JF, D'Cotta H, and Saillant E. 2009. Environmental effects on fish sex determination and differentiation. *Sexual Develop.* 3: 118–35.

Barr CM. 2004. Hybridization and regional sex ratios in *Nemophila menziesii*. *J. Evol. Biol.* 17: 786–94.

Barrett SCH. 1998. The evolution of mating strategies in flowering plants. *Trends Plant Sci.* 3: 335–41.

Barrett SCH. 2002a. The evolution of plant sexual diversity. *Nature Rev. Genet.* 3: 274–84.

Barrett SCH. 2002b. Sexual interference of the floral kind. *Heredity* 88: 154–59.

Barrett SCH, and Charlesworth D. 1991. Effects of a change in the level of inbreeding on the genetic load. *Nature* 352: 522–24.

Bateman AJ. 1948. Intra-sexual selection in *Drosophila*. *Heredity* 2: 349–68.

Bauer RT. 2006. Same sexual system but variable sociobiology: Evolution of protandric simultaneous hermaphroditism in *Lysmata* shrimps. *Int. Comp. Biol.* 46: 430–38.

Bauer RT, and Holt GJ. 1998. Simultaneous hermaphroditism in the marine shrimp *Lysmata wurdemanni* (Caridea: Hippolytidae): An undescribed sexual system in the decapod Crustacea. *Mar. Biol.* 132: 223–35.

Bauer RT, and Newman WA. 2004. Protandric simultaneous hermaphroditism in the marine shrimp *Lysmata californica* (Caridea: Hippolytidae*). J. Crustac. Biol.* 24: 131–39.

Baur B. 1994. Multiple paternity and individual variation in sperm precedence in the simultaneously hermaphroditic land snail, *Arianta arbustorum*. *Behav. Ecol. Sociobiol.* 35: 413–21.

Baur B. 1998. Sperm competition in mollusks. In T. R. Birkhead and A. P. Møller, eds., *Sperm Competition and Sexual Selection*, 255–305. London: Academic Press.

Bawa KS. 1980. Evolution of dioecy in flowering plants. *Annu. Rev. Ecol. Syst.* 11: 15–39.

Bawa KS. 1994. Pollinators of tropical dioecious angiosperms: A reassessment? No, not yet. *Amer. J. Bot.* 81: 456–60.

Bawa KS, and Beach JH. 1981. Evolution of sexual systems in flowering plants. *Ann. Missouri Bot. Gard.* 68: 254–74.

Bawa KS, and Opler PA. 1975. Dioecism in tropical forest trees. *Evolution* 29: 167–79.

Beaumont AT, and Budd MD. 1983. Effects of self-fertilization and other factors on the early development of the scallop *Pecten maximus*. *Marine Biol.* 76: 2885–89.

Bell G. 1982. *The Masterpiece of Nature—The Evolution and Genetics of Sexuality.* Berkeley: U of California P.

Berglund A. 1986. Sex change by a polychaete: Effects of social and reproductive costs. *Ecology* 67: 837–45.

Bergström BI. 1997. Do pandalid shrimp have environmental sex determination? *Marine Biol.* 128: 399–407.

Bertin RI. 1993. Incidence of monoecy and dichogamy in relation to self-fertilization in angiosperms. *Amer. J. Bot.* 80: 557–60.

Bertin RI, and Kerwin MA. 1998. Floral sex ratios and gynomonoecy in *Aster* (Asteraceae). *Amer. J. Bot.* 85: 235–44.

Bertin RI, and Newman CM. 1993. Dichogamy in the angiosperms. *Bot. Rev.* 59: 112–52.

Bhardwaj M, and Eckert CG. 2001. Functional analysis of synchronous dichogamy in flowering rush, *Butomus umbellatus* (Butomaceae). *Amer. J. Bot.* 88: 2204–13.

Bierzychudek P. 1982. The demography of jack-in-the-pulpit, a forest perennial that changes sex. *Ecol. Monogr.* 52: 335–51.

Bierzychudek P. 1984. Assessing "optimal" life histories in a fluctuating environment: The evolution of sex-changing by jack-in-the-pulpit. *Amer. Natur.* 123: 829–40.

Birkhead TR, and Møller AP. 1993a. Female control of paternity. *Trends Ecol. Evol.* 8: 100–104.

Birkhead TR, and Møller AP. 1993b. Sexual selection and the temporal separation of reproductive events: Sperm storage data from reptiles, birds and mammals. *Biol. J. Linnean Soc.* 50: 295–311.

Birkhead TR, and Møller AP, eds. 1998. *Sperm Competition and Sexual Selection.* New York: Academic Press.

Bishop JDD, and Pemberton AJ. 2006. The third way: Spermcast mating mechanism in sessile marine invetebrates. *Integr. Comp. Biol.* 46: 398–406.

Blaber SJM, Brewer DT, Miton DA, Merta GS, Efizon D, Fry G, and van der Velde T. 1999. The life history of the protandrous tropical shad *Tenualosa macrura*

(Alosinae, Clupeidae): Fishery implications. *Estuarine, Coastal, Shelf Sci.* 49: 689–701.

Bond WJ. 1994. Do mutualisms matter? Assessing the impact of pollinator and disperser disruption on plant extinction. *Phil. Trans. Royal Soc. London B* 344: 83–90.

Bonduriansky R. 2009. Reappraising sexual coevolution and the sex roles. *PLoS Biol.* 7: 1–3.

Borgia G, and Blick J. 1981. Sexual competition and the evolution of hermaphroditism. *J. Theoret. Biol.* 89: 523–32.

Branch GM, and Odendaal F. 2003. The effecs of marine protected areas on the population dynamics of a South African limpet, *Cymbula oculus*, relative to the influence of wave action. *Biol. Conserv.* 114: 255–69.

Brauer VS, Schärer L, and Michiels NK. 2007. Phenotypically flexible sex allocation in a simultaneous hermaphrodite. *Evolution* 61: 216–22.

Breder CM, and Rosen DE. 1966. *Modes of Reproduction in Fishes.* Garden City, N.J.: Natural History Press.

Brennan AC, Harris SA, and Hiscock SJ. 2005. Modes and rates of selfing and associated inbreeding depression in the self-incompatible plant *Senecio squalidus* (Asteraceae): A successful colonizing species in the British Isles. *New Phytol.* 168: 475–86.

Brennan AC, Harris SA, and Hiscock SJ. 2006. The population genetics of sporophytic self-compatibility in *Senecio squalidus* L. (Asteraceae): S allele diversity across the British range. *Evolution* 60: 213–24.

Brook HJ, Rawlings TA, and Davies RW. 1994. Protogynous sex change in the intertidal isopod *Gnorimosphaeroma oregonense* (Crustacea: Isopoda). *Biol. Bull.* 187: 99–111.

Brown AHD. 1989. Genetic characterization of plant mating systems. In A. H. D. Brown, M. T. Clegg, A. L. Kahler, and B. S. Weir, eds., *Plant Population Genetics, Breeding, and Genetic Resources*, 145–62. Sunderland, Mass.: Sinauer.

Bruce RW. 1980. Protogynous hermaphroditism in two marine angelfishes. *Copeia* 1980: 353–55.

Brunet J. 1992. Sex allocation in hermaphroditic plants. *Trends Ecol. Evol.* 7: 79–84.

Brusle J, and Brusle S. 1974. Ovarian and testicular intersexuality in two protogynous Mediterranean groupers, *Epinephelus aeneus* and *Epinephelus gauza*. In R. Reinboth, ed., *Intersexuality in the Animal Kingdom*, 222–27. New York: Springer-Verlag.

Bryant EH, and Reed DH. 1999. Fitness decline under relaxed selection in captive populations. *Conserv. Biol.* 13: 665–69.

Budar F, Touzet P, and De Paepe R. 2003. The nucleo-mitochondrial conflict in cytoplasmic male sterilities revisited. *Genetica* 117: 3–16.

Bull JJ. 1983. *Evolution of Sex-Determining Mechanisms.* Menlo Park, Calif.: Benjamin/Cummings.

Bull JJ, and Charnov EL. 1985. On irreversible evolution. *Evolution* 39: 1149–55.

Burt A, and Trivers R. 2006. *Genes in Conflict: The Biology of Selfish Genetic Elements*. Cambridge: Belknap Press.

Busch JW. 2005. The evolution of self-incompatibility in geographically peripheral populations of *Leavenworthia alabamica* (Brassicaceae). *Amer. J. Bot.* 92: 1503–1512.

Buxton CD, and Clarke JR. 1991. The biology of the white musselcracker *Sparodon durbanensis* (Pisces, Sparidae) on the eastern cape coast, South Africa. *S. Afr. J. Marine Sci.* 10: 285–96.

Cahalan CM, and Gliddon C. 1985. Genetic neighborhood sizes in *Primula vulgaris*. *Heredity* 54: 65–70.

Calado R, Bartilotti C, and Narciso L. 2005. Short report on the effect of a parasitic isopod on the reproductive performance of a shrimp. *J. Exptl. Marine Biol. Ecol.* 321: 13–18.

Campbell DR. 1989. Measurements of selection in a hermaphroditic plant: Variation in male and female pollination success. *Evolution* 43: 318–34.

Campbell DR. 1998. Variation in lifetime male fitness in *Ipomopsis aggregata*: Tests of sex allocation theory. *Amer. Natur.* 152: 338–53.

Campbell DR. 2000. Experimental tests of sex-allocation theory in plants. *Trends Ecol. Evol.* 15: 227–32.

Carlon DB. 1999. The evolution of mating systems in tropical coral reefs. *Trends Ecol. Evol.* 14: 491–95.

Carr DE, and Dudash MR. 2003. Recent approaches into the genetic basis of inbreeding depression in plants. *Phil. Trans. Royal Soc. London B* 358: 1071–1084.

Carr GD, Powell EA, and Kyhos DW. 1986. Self-incompatibility in the Hawaiian Madiinae (Compositae): An exception to Baker's rule. *Evolution* 40: 430–34.

Castagna M, and Kraeuter JN. 1994. Age, growth-rate, sexual dimorphism and fecundity of knobbed whelk *Busycon carica* (Gmelin, 1791) in a western mid-Atlantic lagooon system, Virginia. *J. Shellfish Res.* 13: 581–85.

Castric V, and Vekemans X. 2004. Plant self-incompatibility in natural populations: A critical assessment of recent theoretical and empirical advances. *Mol. Ecol.* 13: 2873–89.

Chapman RW, Sedberry GR, Koenig CC, and Eleby BM. 1999. Stock identification of gag, *Mycteroperca microlepis*, along the southeast coast of the United States. *Marine Biotech.* 1: 137–46.

Chapman T, Arnqvist G, Bangham J, and Rowe L. 2003. Sexual conflict. *Trends Ecol. Evol.* 18: 41–47.

Charlesworth B. 1980. The cost of sex in relation to mating system. *J. Theoret. Biol.* 84: 655–71.

Charlesworth B. 1991. The evolution of sex chromosomes. *Science* 251: 1030–1033.

Charlesworth B. 2009. Effective population size and patterns of molecular evolution and variation. *Nature Rev. Genet.* 10: 195–205.

Charlesworth B, and Charlesworth D. 1978. A model for the evolution of dioecy from gynodioecy. *Amer. Natur.* 112: 975–97.

Charlesworth D. 1984. Androdioecy and the evolution of dioecy. *Biol. J. Linnean Soc.* 23: 333–48.

Charlesworth D. 1985. Distribution of dioecy and self-incompatibility in angiosperms. In P. J. Greenwood and Slatkin, M. eds., *Evolution—Essays in Honour of John Maynard Smith*, 237–68. Cambridge: Cambridge UP.

Charlesworth D. 1993. Why are unisexual flowers associates with wind pollination and unspecialized pollinators? *Amer. Natur.* 141: 481–90.

Charlesworth, D. 2002. Plant sex determination and sex chromosomes. *Heredity* 88: 94–101.

Charlesworth D, and Charlesworth B. 1979. The evolutionary genetics of sexual systems in flowering plants. *Proc. Roy. Soc. London B* 205: 513–30.

Charlesworth D, and Charlesworth B. 1981. Allocation of resources to male and female functions in hermaphrodites. *Biol. J. Linnean Soc.* 15: 57–74.

Charlesworth D, and Charlesworth B. 1987a. The effect of investment in attractive structures on allocation to male and female functions in plants. *Evolution* 41: 948–68.

Charlesworth D, and Charlesworth B. 1987b. Inbreeding depression and its evolutionary consequences. *Annu. Rev. Ecol. Syst.* 18: 237–68.

Charlesworth D, and Charlesworth B. 1990. Inbreeding depression with heterozygote advantage and its effect on selection for modifiers changing the outcrossing rate. *Evolution* 44: 870–88.

Charlesworth D, Charlesworth B, and Marais G. 2005. Steps in the evolution of heteromorphic sex chromosomes. *Heredity* 95: 118–28.

Charlesworth D, and Ganders FR. 1979. The population genetics of gynodioecy with cytoplasmic-genic male-sterility. *Heredity* 43: 213–18.

Charlesworth D, and Morgan MT. 1991. Allocation of resources to sex functions in flowering plants. *Phil. Trans. Royal Soc. London B* 332: 91–102.

Charlesworth D, Morgan MT, and Charlesworth B. 1990. Inbreeding depression, genetic load and the evolution of outcrossing rates in a multi-locus system with no linkage. *Evolution* 44: 1469–89.

Charlesworth D, Morgan MT, and Charlesworth B. 1992. The effect of linkage and population size on inbreeding depression due to mutational load. *Genet. Res.* 59: 49–61.

Charnov EL. 1979. Simultaneous hermaphroditism and sexual selection. *Proc. Natl. Acad. Sci. USA* 76: 2480–84.

Charnov EL. 1982. *The Theory of Sex Allocation*, Princeton: Princeton UP.

Charnov EL. 1987a. On sex allocation and selfing in higher plants. *Evol. Ecol.* 1: 30–36.

Charnov EL. 1987b. Sexuality and hermaphroditism in barnacles: A natural selection approach. *Crust. Issues* 5: 89–103.

Charnov EL. 1993. *Life History Invariants*. Oxford: Oxford UP.

Charnov EL. 1996. Sperm competition and sex allocation in simultaneous hermaphrodites. *Evol. Ecol.* 10: 457–62.

Charnov EL, and Bull J. 1977. When is sex environmentally determined? *Nature* 266: 828–30.

Charnov EL, and Hannah RW. 2002. Shrimp adjust their sex ratio to fluctuating age distributions. *Evol. Ecol. Res.* 4: 239–46.

Charnov EL, Maynard Smith J, and Bull JJ. 1976. Why be an hermaphrodite? *Nature* 263: 125–26.

Chase R, and Blanchard KC. 2006. The snail's love-dart delivers mucus to increase paternity. *Proc. Roy. Soc. London B* 273: 1471–75.

Chase R, and Vaga K. 2006. Independence, not conflict, characterizes dart-shooting and sperm exchange in a hermaphroditic snail. *Behav. Ecol. Sociobiol.* 59: 732–39.

Chen M-H, Soong K, and Min-Li T. 2004. Host effect on size structure and timing of sex change in the coral-inhabiting snail *Coralliophila violacea*. *Marine Biol.* 144: 287–93.

Chen X. 1993. Comparison of inbreeding and outbreeding in hermaphroditic *Arianta arbustorum* (L.) (land snail). *Heredity* 71: 456–61.

Choat JH, and Roberson DR. 1974. Protogynous hermaphroditism in fishes of the family Scaridae. In R. Rienboth, ed., *Intersexuality in the Animal Kingdom*, 263–83. New York: Springer-Verlag.

Chornesky EA, and Peters EC. 1987. Sexual reproduction and colony growth in the scleractinian coral *Porites astreoides*. *Biol. Bull.* 172: 161–77.

Clark AB. 1978a. Hermaphroditism as a reproductive strategy for metazoans: Some correlated benefits. *New Zeal. J. Zool.* 5: 769–80.

Clark AB. 1978b. Sex ratio and local resource competition in a prosimian primate. *Science* 201: 163–65.

Clegg MT. 1980. Measuring plant mating systems. *BioScience* 30: 814–18.

Clegg MT, and Allard RW. 1972. Patterns of genetic differentiation in the slender wild oat species *Avena barbata*. *Proc. Natl. Acad. Sci. USA* 69: 1820–24.

Clegg MT, and Allard RW. 1973. Viability versus fecundity selection in the slender wild oat, *Avena barbata* L. *Science* 181: 667–68.

Clegg MT, Allard RW, and Kahler AL. 1972. Is the gene the unit of selection? Evidence from two experimental plant populations. *Proc. Natl. Acad. Sci. USA* 69: 2474–78.

Clout MN, Elliott GP, and Robertson BC. 2002. Effects of supplementary feeding on the offspring sex ratio of kakapo: A dilemma for the conservation of a polygynous parrot. *Biol. Conserv.* 107: 13–18.

Clutton-Brock TH. 2004. What is sexual selection? In P. Kappeler and C. Van Schaik, eds., *Sexual Selection in Primates*, 24–36. Cambridge: Cambridge UP.

Clutton-Brock TH, and Parker GA. 1992. Potential reproductive rates and the operation of sexual selection. *Quart. Rev. Biol.* 67: 437–56.

Clutton-Brock TH, and Vincent ACJ. 1991. Sexual selection and the potential reproductive rates of males and females. *Nature* 351: 58–60.

Cnaani A, and 14 others. 2007. Genetics of sex determination in tilapine species. *Sexual Develop.* 2: 43–54.

Coe WR. 1943. Sexual differentiation in molluscs. *Q. Rev. Biol.* 18: 154–64.

Cohen CS. 1996. The effects of contrasting modes of fertilization on levels of inbreeding in the marine invertebrate genus *Corella. Evolution* 50: 1896–1907.

Cole KS. 1990. Patterns of gonad structure in hermaphroditic gobies (Teleostei: Gobiidae). *Environ. Biol. Fish.* 28: 125–42.

Cole KS, and Noakes DLG. 1997. Gonadal development and sexual allocation in mangrove killifish, *Rivulus marmoratus* (Pisces: Atherinomorpha). *Copeia* 1997: 596–600.

Cole KS, and Shapiro DY. 1995. Social facilitation and sensory mediation of adult sex change in a cryptic, benthic marine goby. *J. Exptl. Marine Biol. Ecol.* 186: 65–75.

Collin R. 1995. Sex, size, and position: A test of models predicting size at sex change in the protandrous gastropod *Crepidula fornicta. Amer. Natur.* 146: 815–31.

Collin R. 2006. Sex ratio, life-history invariants, and patterns of sex change in a family of protandrous gastropods. *Evolution* 60: 735–45.

Collin R, McLellan M, Gruber K, and Bailey-Jourdain C. 2005. Effects of conspecific associations on size at sex change in three species of calyptraeid gastropods. *Marine Ecol. Progr. Ser.* 293: 89–97.

Conn JS, Wentworth TR, and Blum U. 1980. Patterns of dioecism in the flora of the Carolinas. *Amer. Midl. Natur.* 103: 310–15.

Conover DO. 1984. Adaptive significance of temperature-dependent sex-determination in a fish. *Amer. Natur.* 123: 297–313.

Conover DO, and Heins SW. 1987. Adaptive variation in environmental and genetic sex determination in a fish. *Nature* 326: 496–98.

Conover DO, and Kynard BE. 1981. Environmental sex determination: Interaction of temperature and genotype in a fish. *Science* 250: 577–79.

Cooper CG, Miller PL, and Holland PWH. 1996. Molecular genetic analysis of sperm competition in the damselfly *Ischnura elegans* (Vander Linden). *Proc. Roy. Soc. Lond.* B 263: 1343–49.

Costa WJEM. 2004. *Kryptolebias*, a substitute name for *Cryptolebias* Costa, 2004 and Kryptolebiatinae, a substitute name for Cryptolebiatinae Costa, 2004 (Cyprinodontiformes: Rivulidae). *Neotrop Ichthyol* 2: 107–108.

Cox PA. 1982. Vertebrate pollination and the maintenance of dioecy *in Freycinetia. Amer. Natur.* 120: 65–80.

Craig JK, Foote CJ, and Wood CC. 1996. Evidence for temperature-dependent sex determination in sockeye salmon (*Oncorhynchus nerka*). *Canad. J. Fish. Aquat. Sci.* 53: 141–47.

Crisp DJ. 1983. *Chelonobia patula* (Ranzani), a pointer to the evolution of the complemental male. *Marine Biol. Lett.* 4: 281–94.

Crnokrak P, and Roff DA. 1999. Inbreeding depression in the wild. *Heredity* 83: 260–70.

Crow JF. 1954. Breeding structure of populations. II. Effective population number. In T. A. Bancroft, J. W. Gowen, and J. L. Lush, eds., *Statistics and Mathematics in Biology*, 543–56. Ames: Iowa State College Press.

Cruden RW. 1977. Pollen-ovule ratios: A conservative indicator of breeding systems in flowering plants. *Evolution* 31: 32–46.

Cruden RW, and Lyon DL. 1985. Patterns of biomass allocation to male and female functions in plants with different mating systems. *Oecologia* 66: 299–306.

Cruzan MB. 1990. Variation in pollen size, fertilization ability, and postfertilization siring ability in *Erythronium grandiflorum*. *Evolution* 44: 505–515.

Culley TM, and Klooster MR. 2007. The cleistogamous breeding system: A review of its frequency, evolution, and ecology in angiosperms. *Bot. Rev.* 73: 1–30.

Cutter AD, Aviles L, and Ward S. 2003. The proximate determinants of sex ratio in *C. elegans* populations. *Genet. Res.* 81: 91–102.

Cutter AD, and Payseur BA. 2003. Rates of deleterious mutation and the evolution of sex in *Caenorhabditis*. *J. Evol. Biol.* 16: 812–22.

Darwin C. 1851. *A Monograph of the Sub-class Cirripedia*. Vol. 1. London: The Ray Society.

Darwin C. 1854. *A Monograph of the Sub-class Cirripedia*. Vol. 2. London: The Ray Society.

Darwin C. 1859. *On the Origin of Species by Means of Natural Selection*. London: Murray.

Darwin C. 1862a. On the two forms, or dimorphic condition, in the species of *Primula*, and on their remarkable sexual relations. *J. Proc. Linnean Soc. (Botany)* 6: 77–96.

Darwin C. 1862b. *On the Various Contrivances by which British and Foreign Orchids are Fertilized by Insects, and on the Good Effects of Intercrossing*. London: Murray.

Darwin C. 1871. *The Descent of Man, and Selection in Relation to Sex*. London: Murray.

Darwin C. 1876. *The Effects of Cross and Self-fertilization in the Vegetable Kingdom*. London: Murray.

Darwin C. 1877. *The Different Forms of Flowers on Plants of the Same Species*. London: Murray.

Darwin C. 1881. *The Formation of Vegetable Mould, Through the Action of Worms, with Observations on Their Habits*. London: Murray.

Davis WP, Taylor DS, and Turner BJ. 1990. Field observations of the ecology and habits of mangrove Rivulus (*Rivulus marmoratus*) in Belize and Florida (Teleostei: Cyprinodontiformes: Rivulidae). *Icthyol. Explorations Freshw.* 1: 123-34.

Day T, and Aarssen LW. 1997. A time commitment hypothesis for size-dependent gender allocation. *Evolution* 51: 988–93.

de Jong TJ, and Klinkhamer PGL. 1994. Plant size and reproductive success through female and male function. *J. Ecol.* 82: 399–402.

de Jong TJ, Klinkhamer PGL, and Rademaker MCJ. 1999. How geitonogamous selfing affects sex allocation in hermaphroditic plants. *J. Evol. Biol.* 12: 166–76.

de Jong TJ, Waser NM, and Klinkhamer PGL. 1993. Geitonogamy: The neglected side of selfing. *Trends Ecol. Evol.* 8: 321–25.

Dellaporta SL, and Calderon-Urrea A. 1993. Sex determination in flowering plants. *The Plant Cell* 5: 1241–51.

Delph LF. 2003. Sexual dimorphism in gender plasticity and its consequences for breeding system evolution. *Evol. Develop.* 5: 34–39.

Delph LF, and Wolf DE. 2005. Evolutionary consequences of gender plasticity in genetically dimorphic breeding systems. *New Phytol.* 166: 119–28.

de Nettancourt D. 1977. Incompatibility in angiosperms. *Sex. Plant Reprod.* 10: 185–99.

Desfeux C, Maurice S, Henry JP, Lejeune B, and Gouyon PH. 1996. Evolution of reproductive systems in the genus *Silene*. *Proc. Roy. Soc. London B* 263: 409–414.

Desperz D, and Melard C. 1998. Effect of ambient water temperature on sex determination in the blue tilapia *Oreochromis aureus*. *Aquaculture* 162: 1–2.

DeVisser JA, Ter Maat T, and Zonneveld C. 1994. Energy budgets and reproductive allocation in the simultaneous hermaphrodite pond snail, *Lymnaea stagnalis* (L.): A trade-off between male and female function. *Amer. Natur.* 144: 861–67.

Devlin B, and Stephenson AG. 1987. Sexual variations among plants of a perfect-flowered species. *Amer. Natur.* 130: 199–218.

Devlin RH, and Nagahama Y. 2002. Sex determination and sex differentiation in fish: An overview of genetic, physiological, and environmental influences. *Aquaculture* 208: 191–364.

DeWoody JA, Hale MC, and Avise JC. 2010. Vertebrate sex-determining genes and their potential utility in conservation, with particular emphasis on fishes. In J. A. DeWoody, J. W. Bickham, C. Michler, K. Nichols, O. E. Rhodes, and K. Woeste, eds., *Molecular Approaches in Natural Resource Conservation*, 74–99. Cambridge: Cambridge UP.

Doebeli M, and Hauert C. 2005. Models of cooperation based on the Prisoner's Dilemma and the Snowdrift game. *Ecol. Letters* 8: 748–66.

Dolgin ES, Charlesworth B, Baird SE, and Cutter AD. 2007. Inbreeding and outbreeding depression in *Caenorhabditis* nematodes. *Evolution* 61: 1339–52.

Donoghue MJ. 1989. Phylogenies and the analysis of evolutionary sequences, with examples from seed plants. *Evolution* 43: 1137–56.

Dorken ME, and Mitchard ETA. 2008. Phenotypic plasticity of hermaphrodite sex allocation promotes the evolution of separate sexes: An experimental test of the sex-differential plasticity hypothesis using *Sagittaria latifolia* (Alismataceae). *Evolution* 62: 971–78.

Doums C, Viard F, and Jarne P. 1998. The evolution of phally polymorphism. *Biol. J. Linnean Soc.* 64: 273–96.

Doums C, Viard F, Pernot A-F, Delay B, and Jarne P. 1996. Inbreeding depression, neutral polymorphism, and copulatory behavior in freshwater snails: A self-fertilization syndrome. *Evolution* 50: 1908–1918.

Dreger AD. 2001. *Hermaphrodites and the Medical Invention of Sex*. Cambridge: Harvard UP.

Dreger AD, ed. 1999. *Intersex in the Age of Ethics*. College Park, Md.: University Publishing Group.

D'Souza TG, Storhas M, Schulenburg H, Beukeboom LW, and Michiels NK. 2004. Occasional sex in an "asexual" polyploid hermaphrodite. *Proc. Roy. Soc. London B* 271: 1001–1007.

Dudash MR. 1990. Relative fitness of selfed and outcrossed progeny in a self-incompatible, protandrous species, *Sabatia angularis* L (Gentianaceae): A comparison in three environments. *Evolution* 44: 1129–39.

Dunn DF. 1975. Gynodioecy in an animal. *Nature* 253: 528–29.

Dwyer KG, Balent MA, Nasrallah JB, and Nasrallah ME. 1991. DNA sequences of self-incompatibility genes from *Brassica campestris* and *B. oleraceae*: Polymorphism predating speciation. *Plant Mol. Biol.* 16: 481–86.

Eberhard WG. 1985. *Sexual Selection and Animal Genitalia*. Cambridge: Harvard UP.

Eberhard WG. 1998. Female roles in sperm competition. In T. R. Birkhead and A. P. Møller, eds., *Sperm Competition and Sexual Selection*, 91–116. London: Academic Press.

Ebisawa A. 1990. Reproductive biology of *Lethrinus nebulosus* (Pisces, Lethrinidae) around Okinawan waters. *Nippon Suisan Gakkaishi* 56: 1941–54.

Edwardson JR. 1970. Cytoplasmic male sterility. *Bot. Rev.* 36: 341–420.

Elle E, and Meagher TR. 2000. Sex allocation and reproductive success in the andromonoecious perennial *Solanum carolinense* (Solanaceae). II. Paternity and functional gender. *Amer. Natur.* 156: 622–36.

Emlen ST. 1997. When mothers prefer daughters over sons. *Trends Ecol. Evol.* 12: 291–92.

Emlen ST, and Oring LW. 1977. Ecology, sexual selection, and the evolution of mating systems. *Science* 197: 215–23.

Emms SK, Hodges SA, and Arnold ML. 1996. Pollen-tube competition, siring success, and consistent asymmetric hybridization in Louisiana irises. *Evolution* 50: 2201–2206.

Eppley SM, and Jesson LK. 2008. Moving to mate: The evolution of separate and combined sexes in multicellular organisms. *J. Evol. Biol.* 21: 727–36.

Eppley SM, and Pannell JR. 2007. Density-dependent self-fertilization and male versus hermaphrodite siring success in an androdioecious species. *Evolution* 61: 2349–59.

Evanno G, Madec L, and Arnaud J-F. 2005. Multiple paternity and postcopulatory sexual selection in a hermaphrodite: What influences sperm precedence in the garden snail *Helix aspersa*? *Mol. Ecol.* 14: 805–812.

Evans TJ. 1953. The alimentary and vascular systems of *Alderia modesta* (Lovén) in relation to its ecology. *Proc. Malacol. Soc. London* 29: 249–58.

Fausto-Sterling, A. 2000. *Sexing the Body: Gender Politics and the Construction of Sexuality*. New York: Basic Books.

Fausto-Sterling, A. 2004. The five sexes. *J. Politics, Gender, and Culture* 3: 191–205.

Fishelson L. 1970. Protogynous sex reversal in the fish *Anthias squamipinnis* (Teleostei, Anthiidae), regulated by the presence or absence of a male fish. *Nature* 227: 90–91.

Fishelson L, and Galil BS. 2001. Gonad structure and reproductive cycle in the deep-sea hermaphroditic tripodfish *Bathypterois mediterraneus* (Chlorophthalmidae, Teleostei). *Copeia* 2001: 556–60.

Fisher EA. 1980. The relationship between mating system and simultaneous hermaphroditism in the coral reef fish, *Hypoplectrus nigricans* (Serranidae). *Anim. Behav.* 28: 620–33.

Fischer EA. 1981. Sexual allocation in a simultaneously hermaphroditic coral reef fish. *Amer. Natur.* 117: 64–82.

Fischer EA. 1984. Local mate competition and sex allocation in simultaneous hermaphrodites. *Amer. Natur.* 124: 590–96.

Fischer EA. 1988. Simultaneous hermaphroditism, tit-for-tat, and the evolutionary stability of social systems. *Ethol. Sociobiol.* 9: 119–36.

Fischer EA, and Petersen CW. 1987. The evolution of sexual patterns in the seabasses. *BioScience* 37: 482–89.

Fisher RA. 1930. *The Genetical Theory of Natural Selection*. Oxford: Clarendon Press.

Fisher RA. 1941. Average excess and average effect of a gene substitution. *Ann. Eugen* 11: 53–63.

Fleming TH, Maurice S, and Hamrick JL. 1998. Geographic variation in the breeding system and the evolutionary stability of trioecy in *Pachycereus pringlei* (Cactaceae). *Evol. Ecol.* 12: 279–89.

Flinn KM. 2006. Reproductive biology of three fern species may contribute to differential colonization success in post-agricultural forests. *Amer. J. Bot.* 93: 1289–94.

Foltz DW, Ochman H, Jones JS, Evangelisti SM, and Selander RK. 1982. Genetic population structure and breeding systems in arionid slugs (Mollusca: Plumonata). *Biol. J. Linnean Soc.* 17: 225–41.

Foltz DW, Ochman H, and Selander RK. 1984. Genetic diversity and breeding systems in terrestrial slugs of the families Limacidae and Arionidae. *Malacologia* 25: 593–605.

Foster JW, and Marshall-Graves JA. 1994. An Sry-related sequence on the marsupial X-chromosome—Implications for the evolution of the mammalian testis determining gene. *Proc. Natl. Acad. Sci. USA* 91: 1927–31.

Fox JF. 1985. Incidence of dioecy in relation to growth form, pollination and dispersal. *Oecologia* 67: 244–49.

Francis RC. 1992. Sexual lability in teleosts: Developmental factors. *Quart. Rev. Biol.* 67: 1–18.

Frank SA. 2002. A touchstone in the study of adaptation. *Evolution* 56: 2561–64.

Frankham R. 1995. Inbreeding and extinction: A threshold effect. *Conserv. Biol.* 9: 792–99.

Frankham R, Ballou JD, and Briscoe DA. 2002. *Introduction to Conservation Genetics.* Cambridge: Cambridge UP.

Franklin-Tong VE. 2008. *Self-incompatibility in flowering plants: Evolution, Diversity, and Mechanisms.* Berlin: Springer-Verlag.

Freeman DC, Harper KT, and Charnov JG. 1980a. Sex change in plants: Old and new observations and new hypotheses. *Oecologia* 47: 222–32.

Freeman DC, Harper KT, and Ostler WK. 1980b. Ecology of plant dioecy in the intermountain region of western North America and California. *Oecologia* 44: 410–17.

Fricke H, and Fricke S. 1977. Monogamy and sex change by aggressive dominance in coral reef fish. *Nature* 266: 830–32.

Fridolfsson, A. K. and 9 others. 1998. Evolution of the avian sex chromosomes from an ancestral pair of autosomes. *Proc. Natl. Acad. Sci. USA* 95: 8147–52.

Fritsch R, and Rieseberg LH. 1992. High outcrossing rates maintain male and hermaphroditic individuals in populations of the flowering plant *Datisca glomerata. Nature* 359: 633–36.

Fujioka Y. 2001. Thermolabile sex determination in honmoroko. *Fish. Sci.* 68: 889–93.

Garcia-Robledo C, and Mora F. 2007. Pollination biology and the impact of floral display, pollen donors, and distyly on seed production in *Arcytophyllum lavarum* (Rubiaceae). *Plant Biol.* 9: 453–61.

García-Velazco H, Obregón-Barboza H, Rodriguez-Jaramillo C, and Maeda-Martínez AM. 2009. Reproduction of the tadpole shrimp *Triops* (Notostraca) in Mexican waters. *Current Sci.* 96: 91–97.

Garratt PA. 1991. Spawning behavior of *Cheimerius nufar* in captivity. *Environ. Biol. Fish.* 31: 345–53.

Ghiselin MT. 1969. The evolution of hermaphroditism among animals. *Quart. Rev. Biol.* 44: 189–208.

Ghiselin MT. 1974. *The Economy of Nature and the Evolution of Sex.* Berkeley: U of California P.

Gibbs PE, Pascoe PL, and Burt GR. 1988. Sex change in the female dog-whelk, *Nucella lapillus*, induced by tributyltin from antifouling paints. *J. Mar. Biol. Assoc. UK* 68: 715–31.

Gilbert SF, and Epel D. 2009. *Ecological Developmental Biology: Integrating Epigenetics, Medicine, and Evolution.* Sunderland, Mass.: Sinauer.

Gillespie JH. 1977. Natural selection for within-generation variance in offspring numbers: A new evolutionary principle. *Amer. Natur.* 111: 1010–1014.

Givnish TJ. 1980. Ecological constraints on the evolution of breeding systems in seed plants: Dioecy and dispersal in gymnosperms. *Evolution* 34: 959–72.

Givnish TJ. 1982. Outcrossing versus ecological constraints in the evolution of dioecy. *Amer. Natur.* 119: 849–65.

Godwin J. 1994. Behavioral aspects of protandrous sex change in the anemonefish *Amphiprion melanopus* and endocrine correlates. *Anim. Behav.* 48: 551–67.

Goldman DA, and Willson MF. 1986. Sex allocation in functionally hermaphroditic plants: A review and critique. *Bot. Rev.* 52: 157–94.

Goodwillie C, Kalisz S, and Eckert CG. 2005. The evolutionary enigma of mixed mating systems in plants: Occurrence, theoretical explanations, and empirical evidence. *Annu. Rev. Ecol. Evol. Syst.* 36: 47–79.

Gouyon PH, and Couvet D. 1988. A conflict between two sexes, females and hermaphrodites. In S. C. Stearns, ed., *The Evolution of Sex and Its Consequences*, 245–61. Basel, Switzerland: Birkhauser Verlag.

Gowaty PA, and Hubbell SP. 2009. Reproductive decisions under ecological constraints: It's about time. *Proc. Natl. Acad. Sci. USA* 106: 10017–10024.

Graff C, Clayton DA, and Larsson N-G. 1999. Mitochondrial medicine—Recent advances. *J. Internal Med.* 246: 11–23.

Greeff JM, and Michiels NK. 1999. Sperm digestion and reciprocal sperm transfer can drive hermaphroditic sex allocation to equality. *Amer. Natur.* 153: 421–30.

Greeff JM, Nason JD, and Compton SG. 2001. Skewed paternity and sex allocation in hermaphroditic plants and animals. *Proc. Roy. Soc. London B* 268: 2143–47.

Griffith SC, Owens IPF, and Thuman KA. 2002. Extra pair paternity in birds: A review of interspecific variation and adaptive function. *Molec. Ecol.* 11: 2195–2212.

Grützner F, and 7 others. 2004. In the platypus a meiotic chain of ten sex chromosomes shares genes with the bird Z and mammal X chromosomes. *Nature* 432: 913–17.

Guo X, and Allen SK Jr. 1994. Sex determination and polyploid gigantism in the dwarf surfclam (*Mulinia lateralis* Say). *Genetics* 138: 1199–1206.

Guo X, Hedgecock D, Hershberger WK, Cooper K, and Allen SK Jr. 1998. Genetic determinants of protandric sex in the Pacific oyster, *Crassostrea gigas* Thunberg. *Evolution* 52: 394–402.

Guo Y, Lang S, and Ellis RE. 2009. Independent recruitment of F box genes to regulate hermaphrodite development during nematode evolution. *Curr. Biol.* 19: 1853–60.

Gwynne DT. 1991. Sexual competition among females: What causes courtship-role reversal? *Trends Ecol. Evol.* 6: 118–21.

Gwynne DT, and Simmons LW. 1990. Experimental reversal of courtship roles in an insect. *Nature* 346: 172–74.

Haag ES. 2005. The evolution of nematode sex determination: *C. elegans* as a reference point for comparative biology. In The *C. elegans* Research Community, ed., *Wormbook*_doi/10.1895/wormbook.1.120.1 (*see* www.wormbook .org).

Haag ES. 2009. Convergent evolution: Regulatory lightning strikes twice. *Curr. Biol.* 19: R977–78.

Haley LE. 1977. Sex determination in the American oyster. *J. Heredity* 68: 114–16.

Haley LE. 1979. Genetics of sex determination in the American oyster. *Proc. Natl. Shellfish. Assoc.* 69: 54–57.

Hall VR, and Hughes TP. 1996. Reproductive strategies of modular organisms: Comparative studies of reef-building corals. *Ecology* 77: 950–63.

Hamilton WD. 1967. Extraordinary sex ratios. *Science* 156: 477–88.

Hamrick JL, and Allard RW. 1972. Microgeographical variation in allozyme frequencies in *Avena barbata*. *Proc. Natl. Acad. Sci. USA* 69: 2100–2104.

Hanson MR. 1991. Plant mitochondrial mutations and male sterility. *Annu. Rev. Genet.* 25: 461–86.

Harder LD, and Barrett SCH. 1996. Pollen dispersal and mating patterns in animal-pollinated plants. In D. G. Lloyd and S. C. H. Barrett, eds., *Floral Biology: Studies on Floral Evolution in Animal-pollinated Plants*, 140–90. New York: Chapman & Hall.

Harder LD, Barrett SCH, and Cole WW. 2000. The mating consequences of sexual segregation within inflorescences of flowering plants. *Proc. Roy. Soc. London B* 267: 315–20.

Harder LD, and Wilson WG. 1998. A clarification of pollen discounting and its joint effects with inbreeding depression on mating system evolution. *Amer. Natur.* 152: 684–95.

Hardy ICW, ed. 2002 *Sex Ratios: Concepts and Research Methods*. Cambridge: Cambridge UP.

Harrington RW. 1961. Oviparous hermaphroditic fish with internal self-fertilization. *Science* 134: 1749–50.

Harrington RW, Jr. 1967. Environmentally controlled induction of primary male gonochorists from eggs of the self-fertilizing hermaphroditic fish, *Rivulus marmoratus* Poey. *Biol. Bull.* 132: 174–99.

Harrington RW, Jr. 1968. Delimitation of the thermolabile phenocritical period of sex determination from eggs of the self-fertilizing hermaphroditic fish *Rivulus marmoratus*. *Physiol. Zool.* 41: 447–60.

Harrington RW. 1971. How ecological and genetic factors interact to determine when self-fertilizing hermaphrodites of *Rivulus marmoratus* change into functional secondary males, with a reappraisal of modes of intersexuality among fishes. *Copeia* 1971: 1239–46.

Harrington RW, and Kallman KD. 1967. The homozygosity of clones of the self-fertilizing hermaphroditic fish *Rivulus marmoratus* (Cyprinodontidae, Atheriniformes). *Amer. Natur.* 102: 337–43.

Harrington RW, Jr., and Kallman KD. 1968. The homozygosity of clones of the self-fertilizing hermaphrodite fish *Rivulus marmoratus* Poey (Cyprinodontidae, Atheriniformes). *Amer. Natur.* 102: 337–43.

Harrison PL, and Wallace CC. 1991. Reproduction, dispersal and recruitment of scleractinian corals. In Z. Dubinsky, ed., *Ecosystems of the World (Vol. 25): Coral Reefs*, 133–207. Amsterdam: Elsevier.

Hart JA. 1985a. Peripheral isolation and the origin of diversity in *Lepechinia* sect. *Parviflorae* (Lamiaceae). *Syst. Bot.* 10: 134–46.

Hart JA. 1985b. Evolution of dioecism in *Lepechinia* Willd. sect. *Parviflorae* (Lamiaceae). *Syst. Bot.* 10: 147–54.

Hartl DL, and Clark AG. 1997. *Principles of Population Genetics.* Sunderland, Mass.: Sinauer.

Harvey PH, Leigh Brown AJ, Maynard Smith J, and Nee S, eds. 1996. *New Uses for New Phylogenies.* Oxford: Oxford UP.

Hastings PA, and Petersen CW. 1986. A novel sexual pattern in serranid fishes: Simultaneous hermaphrodites and secondary males in *Serranus fasciatus. Environ. Biol. Fish.* 15: 59–68.

Hattori A. 1991. Socially controlled growth and size-dependent sex-change in the anemonefish *Amphiprion frenatus* in Okinawa, Japan. *Jap. J. Ichthyol.* 38: 165–77.

Heath DJ. 1977. Simultaneous hermaphroditism: cost and benefit. *J. Theoret. Biol.* 64: 363–73.

Heath DJ. 1979. Brooding and the evolution of hermaphroditism. *J. Theoret. Biol.* 81: 151–55.

Hecker M, and 12 others. 2006. Terminology of gonadal anomalies in fish and amphibians resulting from chemical exposures. *Rev. Environ. Contam. Toxicol.* 187: 103–131.

Hedgecock D, Chow V, and Waples RS. 1992. Effective population numbers of shellfish broodstocks estimated from temporal variance in allelic frequencies. *Aquaculture* 108: 215–32.

Hedrick PW. 2000. *Genetics of Populations.* Sudbury, Mass.: Jones & Bartlett.

Heilbuth JC. 2000. Lower species richness in dioecious clades. *Amer. Natur.* 156: 221–41.

Helfman GS, Collette BB, and Facey DE. 1997. *The Diversity of Fishes.* Malden, Mass.: Blackwell.

Heller J. 1993. Hermaphroditism in mollusks. *Biol. J. Linnean Soc.* 48: 19–42.

Hill RC, de Carvalho CE, Salogiannis J, Schlager B, Pilgrim D, and Haag ES. 2006. Genetic flexibility in the convergent evolution of hermaphroditism in *Caenorhabditis* nematodes. *Develop. Cell* 10: 531–38.

Hinck JE, Blazer VS, Schmitt XCJ, Papoulias DM, and Tillett DE. 2009. Widespread occurrence of intersex in black basses (*Micropterus* spp.) from U.S. rivers, 1995–2004. *Aquat. Toxicol.* 95: 60–70.

Hines M. 2003. *Brain Gender.* Cambridge: Cambridge UP.

Hoagland KE. 1978. Protandry and the evolution of environmentally-mediated sex change: A study of the Mollusca. *Malacologia* 17: 365–91.

Hoar WS. 1969. Reproduction. In *Fish Physiology,* vol. 3: *Reproduction and Growth, Bioluminescence, Pigments, and Poisons.* New York: Academic Press.

Hoch JM. 2008. Variation in penis morphology and mating ability in the acorn barnacle, *Semibalanus balanoides. J. Exptl. Marine Biol. Ecol.* 359: 126–30.

Hoch JM. 2009. Adaptive plasticity of the penis in a simultaneous hermaphrodite. *Evolution* 63: 1946–53.

Hodgkin J, and Brenner S. 1977. Mutations causing transformation of sexual phenotype in the nematode *Caenorhabditis elegans. Genetics* 162: 767–80.

Høeg JT. 1995. Sex and the single cirripede: A phylogenetic perspective. In *Crustacean Issues 10. New Frontiers in Barnacle Evolution,* 195–207. Rotterdam: Balkema.

Hoffman SG, Schildhauer MP, and Warner RR. 1985. The costs of changing sex and the ontogeny of males under contest competition for mates. *Evolution* 39: 915–27.

Holsinger KE. 1996. Pollination biology and the evolution of mating systems in flowering plants. In M. K. Hecht, ed., *Evolutionary Biology,* 107–149. New York: Plenum.

Hooper RE, and Siva-Jothy MT. 1996. Last male sperm precedence in a damselfly demonstrated by RAPD profiling. *Mol. Ecol.* 5: 449–52.

Hosken DJ, and Stockley P. 2004. Sexual selection and genital evolution. *Trends Ecol. Evol.* 19: 87–93.

Howarth B. 1974. Sperm storage as a function of the female reproductive tract. In A. D. Johnson and C. E. Foley, eds., *The Oviduct and Its Functions,* 237–70. New York: Academic Press.

Hughes RN, Manríquez PH, Bishop JDD, and Burrows MT. 2003. Stress promotes maleness in hermaphroditic modular animals. *Proc. Natl. Acad. Sci. USA* 100: 10326–10330.

Hurst LD. 1990. Parasite diversity and the evolution of diploidy, multicellularity, and anisogamy. *J. Theoret. Biol.* 144: 429–33.

Hurst LD, and Hamilton WD. 1992. Cytoplasmic fusion and the nature of sexes. *Proc. Royal Soc. Lond.* B 247: 189–94.

Husband BC, and Schemske DW. 1996. Evolution and the magnitude and timing of inbreeding depression in plants. *Evolution* 50: 54–70.

Husband BC, and Schemske DW. 1997. The effect of inbreeding in diploid and tetraploid populations of *Epilobium angustifolium* (Onagraceae): Implications for the genetic basis of inbreeding depression. *Evolution* 51: 737–46.

Igléias SP, Sellos DY, and Nakaya K. 2005. Discovery of a normal hermaphroditic chondrichthyan species: *Apristurus longicephalus*. *J. Fish Biol.* 66: 417–28.

Ioerger TR, Clark AG, and Kao T-H. 1990. Polymorphism at the self-incompatibility locus in Solanaceae predates speciation. *Proc. Natl. Acad. Sci. USA* 87: 9732–37.

Irish EE, and Nelson T. 1989. Sex determination in monoecious and dioecious plants. *The Plant Cell* 1: 737–44.

Iwasa Y. 1991. Sex-change evolution and cost of reproduction. *Behav. Ecol.* 2: 56–68.

Jacobs MS, and Wade MJ. 2003. A synthetic review of the theory of gynodioecy. *Amer. Natur.* 161: 837–51.

Jain SK, and Allard RW. 1966. The effects of linkage, epistasis and inbreeding on population changes under selection. *Genetics* 53: 633–59.

Jarne P, and Auld JR. 2006. Animals mix it up too: The distribution of self-fertilization among hermaphroditic animals. *Evolution* 60: 1816–24.

Jarne P, and Charlesworth B. 1993. The evolution of the selfing rate in functionally hermaphroditic plants and animals. *Annu. Rev. Ecol. Syst.* 24: 441–66.

Jarne P, Perdieu M-A, Pernot A-F, Delay B, and David P. 2000. The influence of self-fertilization and grouping on fitness attributes in the freshwater snail *Physa acuta*: population and individual inbreeding depression. *J. Evol. Biol.* 13: 645–55.

Jarne P, Vianey-Liaud M, and Delay B. 1993. Selfing and outcrossing in hermaphroditic freshwater gastropods (Basommatophora): where, when and why. *Biol. J. Linnean Soc.* 49: 99–125.

Jenner MG. 1979. Pseudohermaphroditism in *Ilyanassa obsoleta* (Mollusca: Neogastropoda). *Science* 205: 1407–1409.

Johnston MO, Das B, and Hoeh WR. 1998. Negative correlation between male allocation and rate of self-fertilization in a hermaphroditic animal. *Proc. Natl. Acad. Sci. USA* 95: 617–20.

Jones AG, and Ardren WR. 2003. Methods of parentage analysis in natural populations. *Molec. Ecol.* 12: 2511–23.

Jones AG, Arguello JR, and Arnold SJ. 2002. Validation of Bateman's principles: A genetic study of sexual selection and mating patterns in the rough-skinned newt. *Proc. Royal Soc. London B* 269: 2533–39.

Jones AG, and Avise JC. 2001. Mating systems and sexual selection in male-pregnant pipefishes and seahorses: Insights from microsatellite-based studies of maternity. *J. Heredity* 92: 150–58.

Jones AG, and Ratterman NL. 2009. Mate choice and sexual selection: What have we learned since Darwin? *Proc. Natl. Acad. Sci. USA* 106: 10001–10008.

Jones AG, Rosenqvist G, Berglund A, Arnold SJ, and Avise JC. 2000. The Bateman gradient and the cause of sexual selection in a sex-role-reversed pipefish. *Proc. Royal Soc. London B* 267: 677–80.

Jones AG, Small CM, Paczolt KA, and Ratterman NL. 2009. A practical guide to methods of parentage analysis. *Molec. Ecol. Res.* 10: 6–30.

Jordaens K, Van Dongen S, Temmerman K, and Backeljau T. 2006. Resource allocation in a simultaneously hermaphroditic slug with phally polymorphism. *Evol. Ecol.* 20: 535–48.

Jormalainen V. 1998. Precopulatory mate guarding in crustaceans: Male competitive strategy and intersexual conflict. *Quart. Rev. Biol.* 73: 275–304.

Juarez C, and Banks JA. 1998. Sex determination in plants. *Curr. Opin. Plant Biol.* 1: 68–72.

Just W, and 6 others. 1995. Absence of SRY in species of the vole *Ellobius*. *Nature Genet.* 11: 117–18.

Just W, and 8 others. 2007. *Ellobius lutescens*: Sex determination and sex chromosome. *Sexual Devel.* 1: 211–221.

Kallman KD, and Harrington RW, Jr. 1964. Evidence for the existence of homozygous clones in the self-fertilizing hermaphroditic teleost *Rivulus marmoratus* (Poey). *Biol. Bull.* 126: 101–114.

Kaul MLH. 1988. *Male Sterility in Higher Plants*. Berlin: Springer-Verlag.

Kenchington E, MacDonald B, Cao L, Tsagkarakis D, and Zouros E. 2002. Genetics of mother-dependent sex ratio in blue mussels (*Mytilus* sp.) and implications for doubly uniparental inheritance of mitochondrial DNA. *Genetics* 161: 1579–88.

Kikuchi K, and 6 others. 2007. The sex-determining locus of the tiger pufferfish, *Takifugu rubripes*. *Genetics* 175: 2039–2042.

Killingback T, and Doebeli M. 2002. The continuous prisoner's dilemma and the evolution of cooperation through reciprocal altruism with variable investment. *Amer. Natur.* 160: 421–38.

King AD. 1966. Hermaphroditism in the common dogfish (*Scyliorhinus canicula*). *J. Zool.* 148: 312–14.

Kiontke K, Gavin NP, Raynes Y, Roehrig C, Piano F, and Fitch DHA. 2004. *Caenorhabditis* phylogeny predicts convergence of hermaphroditism and extensive intron loss. *Proc. Natl. Acad. Sci. USA* 101: 9003–9008.

Klinkhamer GL, and de Jong TJ. 2002. Sex allocation in hermaphroditic plants. In I. C. W. Hardy, ed., *Sex Ratios: Concepts and Research Methods*, 333–48. Cambridge: Cambridge UP.

Klinkhamer GL, de Jong TJ, and Metz H. 1997. Sex and size in cosexual plants. *Trends Ecol. Evol.* 12: 260–65.

Knight TM, and 9 others. 2005. Pollen limitation of plant reproduction: Pattern and process. *Annu. Rev. Ecol. Evol. Syst.* 36: 467–97.

Knowlton N, and Greenwell SR. 1984. Male sperm competition avoidance mechanisms: The influence of female interests. In R. L. Smith, ed., *Sperm Competition and the Evolution of Animal Mating Systems*, 61–84. New York: Academic Press.

Knowlton N, and Jackson J. 1993. Inbreeding and outbreeding in marine invertebrates. In N. Thornhill, ed., *The Natural History of Inbreeding and Outbreeding*, 200–249. Chicago: U of Chicago P.

Kobayashi K, and Suzuki K. 1992. Hermaphroditism and sexual function in *Cirrhitichthys aureus* and other Japanese hawkfishes (Cirrhitidae: Teleostei). *Jap. J. Ichthy.* 38: 397–410.

Kobayashi Y, Sunobe T, Kobayashi T, Nagahama Y, and Nakamura M. 2005. Gonadal structure of the serial-sex changing gobiid fish *Trimma okinawae*. *Develop. Growth Differen.* 47: 7–13.

Koene JM. 2006. Tales of two snails: Sexual selection and sexual conflict in *Lymnaea stagnalis* and *Helix aspersa*. *Int. Comp. Biol.* 46: 419–29.

Koene JM, and Chase R. 1998. Changes in the reproductive system of the snail *Helix aspersa* caused by mucus from the love dart. *J. Exptl. Biol.* 201: 2313–19.

Koene JM, Montagne-Wajer K, and Ter Matt A. 2006. Effects of frequent mating on sex allocation in the simultaneously hermaphroditic great pond snail (*Lymnaea stagnalis*). *Behav. Ecol. Sociobiol.* 60: 332–38.

Koene JM, Pförtner T, and Michiels NK. 2005. Piercing the partner's skin influences sperm uptake in the earthworm *Lumbricus terrestris*. *Behav. Ecol. Sociobiol.* 59: 243–49.

Koene JM, and Schulenburg H. 2005. Shooting darts: Co-evolution and counteradaptation in hermaphroditic snails. *BMC Evol. Biol.* 5: 25–38.

Koene JM, Sundermann G, and Michiels NK. 2002. On the function of body piercing during copulation in earthworms. *Invert. Reprod. Develop.* 41: 35–40.

Koene JM, and Ter Maat A. 2002. The distinction between pheromones and allohormones. *J. Comp. Physiol.* 188: 163–64.

Koene JM, and Ter Maat A. 2005. Sex role alternation in the simultaneous hermaphroditic pond snail *Lymnaea stagnalis* is determined by the availability of seminal fluid. *Anim. Behav.* 69: 845–50.

Køie M. 1969. On the endoparasites of *Buccinum undatum* L. with special reference to the trematodes. *Ophelia* 6: 251–79.

Komdeur J, Daan S, Tinbergen J, and Mateman C. 1997. Extreme adaptive modification in sex ratio of the Seychelles warbler's eggs. *Nature* 385: 522–25.

Komdeur J, Magrath MJL, and Krackow S. 2002. Pre-ovulation control of hatchling sex ratio in the Seychelles warbler. *Proc. Royal Soc. London B* 269: 1067–72.

Kristensen H. 1970. Competition in three cyprinodont fish species in the Netherlands Attilles. *Stud. Fauna of Curacao and Other Caribbean Islands* 32: 82-101.

Kuwamura T, Nakashima Y, and Yogo Y. 1994. Sex change in either direction by growth rate advantage in a monogamous coral goby *Paragobiodon echinocephalus*. *Behav. Ecol.* 5: 434–38.

Kuwamura T, and Nakashima Y. 1998. New aspects of sex change among reef fishes: Recent studies in Japan. *Environ. Biol. Fish.* 52: 125–35.

Kuwamura T, Yogo Y, and Nakashima Y. 1993. Size-assortative monogamy and parental egg care in a coral goby *Paragobiodon echniocephalus*. *Ethology* 95: 65–75.

Kvarnemo C, and Ahnesjö I. 1996. The dynamics of operational sex ratios and competition for mates. *Trends Ecol. Evol.* 11: 404–412.

Lagomarsino IV, and Conover DO. 1993. Variation in environmental and genotypic sex-determining mechanisms across a latitudinal gradient in the fish, *Menidia menidia. Evolution* 47: 487–94.

Lahn BT, and Page DC. 1999. Four evolutionary strata on the human X chromosome. *Science* 286: 964–67.

Lande R, Schemske D, and Schultz S. 1994. High inbreeding depression, selective interference among loci, and the threshold selfing rate for purging recessive lethal mutations. *Evolution* 48: 965–78.

Landolfa MA. 2002. On the adaptive function of gamete trading in the black hamlet *Hypoplectrus nigricans. Evol. Ecol. Res.* 4: 1191–99.

Langlois TH. 1965. The conjugal behavior of the introduced giant garden slug, *Limax maximus* L, as observed on South Bass Island, Lake Erie. *Ohio J. Sci.* 65: 298–304.

Laser KD, and Lersten NR. 1972. Anatomy and cytology of microsporogenesis in cytoplasmic male sterile angiosperms. *Bot. Rev.* 38: 425–54.

Lau P, and Bosque C. 2003. Pollen flow in the distylous *Palicourea fendleri* (Rubiaceae): An experimental test of the dissortative pollen flow hypothesis. *Oecologia* 135: 593–600.

Lau PPF, and Sadovy Y. 2001. Gonad structure and sexual pattern in two threadfin breams and possible function for the dorsal accessory duct. *J. Fish Biol.* 58: 1438–53.

Laughlin TL, Lubinski BA, Park EHH, Taylor DS, and Turner BS. 1995. Clonal stability and mutation in the self-fertilizing hermaphroditic fish, *Rivulus marmoratus. J. Heredity* 86: 399–402.

Lee J.-S., Miya M, Lee YS, Kim CG, Park EH, Aoki Y, and Nishida M. 2001. The complete DNA sequence of the mitochondrial genome of the self-fertilizing fish *Rivulus marmoratus* (Cyprinodontiformes, Rivulidae) and the first description of duplication of a control region in fish. *Gene* 280: 1–7.

Leonard JL. 1990. The hermaphrodite's dilemma. *J. Theor. Biol.* 147: 361–71.

Leonard JL. 1992. The "love-dart" in helicid snails: A gift of calcium or a firm commitment? *J. Theoret. Biol.* 159: 513–21.

Leonard JL. 1993. Sexual conflict in simultaneous hermaphrodites: Evidence from serranid fishes. *Environ. Biol. Fish.* 36: 135–48.

Leonard JL. 1999. Modern Portfolio Theory and the prudent hermaphrodite. *Invert. Rep. Dev.* 36: 129–35.

Leonard JL. 2005. Bateman's principle and simultaneous hermaphrodites: A paradox. *Integr. Comp. Biol.* 45: 856–73.

Leonard JL. 2006. Sexual selection: Lessons from hermaphroditic mating systems. *Int. Comp. Biol.* 46: 349–67.

Leonard JL, and Lukowiak K. 1984. Male-female conflict in a simultaneous hermaphrodite resolved by sperm trading. *Amer. Natur.* 124: 282–86.

Leonard JL, and Lukowiak K. 1985. Courtship, copulation and sperm-trading in the sea slug, *Navanax inermis* (Opisthobranchia: Cephalaspidea). *Can. J. Zool.* 63: 2719–29.

Leonard JL, Pearse JS, and Harper AB. 2002. Comparative reproductive biology of *Ariolimax californicus* and *A. dolichophallus* (Gastropoda: Stylommatophora). *Invert. Rep. Dev.* 41: 83–93.

Levitan DR, and Petersen C. 1995. Sperm limitation in the sea. *Trends Ecol. Evol.* 10: 228–31.

Lewis D. 1942. The evolution of sex in flowering plants. *Biol. Rev.* 17: 46–67.

Liston A, Rieseberg LH, and Elias TS. 1990. Functional androdioecy in the flowering plant *Datisca glomerata*. *Nature* 343: 641–42.

Lively CM. 1990. Male allocation and the cost of biparental sex in a parasitic worm. *Lect. Math. Life Sci.* 22: 93–107.

Lloyd DG. 1972. Breeding systems in *Cotula* L. (Compositae, Anthemideae). I. The array of monoclinous and diclinous systems. *New Phytol.* 71: 1181–94.

Lloyd DG. 1975. The maintenance of gynodioecy and androdioecy in angiosperms. *Genetica* 45: 325–39.

Lloyd DG. 1979a. Parental strategies of angiosperms. *New Zeal. J. Bot.* 17: 595–606.

Lloyd DG. 1979b. Some reproductive factors affecting the selection of self-fertilization in plants. *Amer. Natur.* 113: 67–79.

Lloyd DG. 1980. Sexual strategies in plants. III. A quantitative method for describing gender of plants. *New Zeal. J. Bot.* 18: 103–108.

Lloyd DG. 1984. Gender allocation in outcrossing cosexual plants. In R. Dirzo and J. Sarukhan, eds., *Perspectives in Plant Population Ecology*, 277–300. Sunderland, Mass.: Sinauer.

Lloyd DG. 1987a. A general principle for the allocation of limited resources. *Evol. Ecol.* 2:175–87.

Lloyd DG. 1987b. Allocations to pollen, seeds, and pollination mechanisms in self-fertilizing plants. *Funct. Ecol.* 1: 83–89.

Lloyd DG, and Bawa KS. 1984. Modification of the gender of seed plants in varying conditions. *Evol. Biol.* 17: 255–336.

Lloyd DG, and Myall AJ. 1976. Sexual dimorphism in *Cirsium arvense*. *Ann. Bot.* 40: 115–23.

Lloyd DG, and Schoen DJ. 1992. Self- and cross-fertilization in plants. I. Functional dimensions. *Int. J. Plant Sci.* 153: 358–69.

Lloyd DG, and Webb CJ. 1977. Secondary sex characters in plants. *Botan. Rev.* 43: 177–216.

Lloyd DG, and Webb CJ. 1986. The avoidance of interference between the presentation of pollen and stigmas in angiosperms. I. Dichogamy. *New Zealand J. Bot.* 24: 135–62.

Locher R, and Baur B. 2000. Mating frequency and resource allocation in male and female function in the simultaneous hermaphrodite land snail *Arianta arbustorum*. *J. Evol. Biol.* 13: 607–614.

Loewe L, and Cutter AD. 2008. On the potential for extinction by Muller's ratchet in *Caenorhabditis elegans*. *BMC Evol. Biol.* 8: 125–38.

Longhurst AR. 1955 Evolution in the Notostraca. *Evolution* 9: 84–86.

López S, and Domínguez CA. 2003. Sex choice in plants: Facultative adjustment of the sex ratio in the perennial herb *Begonia gracilis*. *J. Evol. Biol.* 16: 1177–85.

Lorch PD, Bussiere LF, and Gwynne DT. 2008. Quantifying the potential for sexual dimorphism using upper limits on Bateman gradients. *Behaviour* 145: 1–24.

Lorenzi MC, Schleicherova D, and Sella G. 2006. Life history and sex allocation in the simultaneously hermaphroditic polychaete worm *Ophryotrocha diadema*: The role of sperm competition. *Integr. Comp. Biol.* 46: 381–89.

Lorenzi MC, Schleicherova D, and Sella G. 2008. Sex adjustments are not functionally costly in simultaneous hermaphrodites. *Marine Biol.* 153: 599–604.

Lorenzi V, Earley RL, and Grober MS. 2006. Preventing behavioral interactions with a male facilitates sex change in female blue banded gobies, *Lythrypnus dalli*. *Behav. Ecol. Sociobiol.* 59: 715–22.

Lovett-Doust L, and Lovett-Doust J. 1988. *Plant Reproductive Ecology: Patterns and Strategies*. New York: Oxford UP.

Lubinski BA, Davis WP, Taylor DS, and Turner BJ. 1995. Outcrossing in a natural population of a self-fertilizing hermaphroditic fish. *J. Heredity* 86: 469–73.

Lutnesky MMF. 1994. Density-dependent protogynous sex change in territorial-haremic fishes: Models and evidence. *Behav. Ecol.* 5: 375–83.

Mack PD, Priest NK, and Promislow DEL. 2003. Female age and sperm competition: Last-male precedence declines as female age increases. *Proc. Roy. Soc. Lond. B* 270: 159–65.

Mackiewicz M., Tatarenkov A, Perry A, Martin JR, Elder, DF, Bechler DL, and Avise JC. 2006a. Microsatellite documentation of male-mediated outcrossing between inbred laboratory strains of the self-fertilizing mangrove killifish (*Kryptolebias marmoratus*). *J. Heredity* 97: 508–513.

Mackiewicz M, Tatarenkov A, Turner BJ, and Avise JC. 2006b. A mixed-mating strategy in a hermaphroditic vertebrate. *Proc. Roy. Soc. Lond. B* 273: 2449–52.

Mackiewicz M., Tatarenkov A, Taylor DS, Turner BJ, and Avise JC. 2006c. Extensive outcrossing and androdioecy in a vertebrate species that otherwise reproduces as a self-fertilizing hermaphrodite. *Proc. Natl. Acad. Sci. USA* 103: 9924–28.

Mair GC, Beardmore JA, and Skibinski DOF. 1980. Experimental evidence for environmental sex determination in *Oreochromis* speces. In R. Hirano and I. Hanyu, eds., *Proceedings of the Second Asian Fisheries Forum*, 555–58. Manila (Philippines): Asian Fisheries Society.

Mair GC, Scott AG, Penman DJ, Beardmore JA, and Skibinski DOF. 1991. Sex determination in the genus *Oreochromis*. I. Sex reversal, gynogenesis and triploidy in *Oreochromis niloticus*. *Theoret. Appl. Genet.* 82: 144–52.

Majerus MEN. 2003. *Sex Wars: Genes, Bacteria, and Biased Sex Ratios*. Princeton: Princeton UP.

Mank JE. 2006. The evolution of reproductive and genomic diversity in ray-finned fishes. Ph.D. diss. Athens: University of Georgia.

Mank JE, and Avise JC. 2009. Evolutionary diversity and turn-over of sex determination in teleost fishes. *Sexual Develop.* 3: 60–67.

Mank JE, and Ellegren H. 2007. Parallel divergence and degradation of the avian W sex chromosome. *Trends Ecol. Evol.* 22: 389–91.

Mank JE., Promislow DEL, and Avise JC. 2006. Evolution of alternative sex-determining mechanisms in teleost fishes. *Biol. J. Linnean Soc.* 87: 83–93.

Margulis, L. 1970. *Origin of Eukaryotic Cells*. New Haven: Yale UP.

Marshall DL, and Folsom MW. 1991. Mate choice in plants: An anatomical to population perspective. *Annu. Rev. Ecol. Syst.* 22: 37–63.

Martin A, and 8 others. 2009. A transposon-induced epigenetic change leads to sex determination in melon. *Nature* 461: 1135–38.

Maruyama T, and Kimura M. 1980. Genetic variability and effective population size when local extinction and recolonization of subpopulations are frequent. *Proc. Natl. Acad. Sci. USA* 77: 6710–14.

Massaro EJ, Massaro JC, and Harrington RW, Jr. 1975. Biochemical comparison of genetically different homozygous clones (isogenic, uniparental lines) of the self-fertilizing fish *Rivulus marmoratus* Poey. In C. L. Markert, ed., *Isozymes*, vol. 3:439–53. New York: Academic Press.

Matton DP, Nass N, Clarke A, and Newbigin E. 1994. Self incompatibility: How plants avoid illegitimate offspring. *Proc. Natl. Acad. Sci. USA* 91: 1992–97.

Maurice S, Charlesworth D, Desfeux C, Couvet D, and Gouyon P-H. 1993. The evolution of gender in hermaphrodites of gynodioecious populations with nucleo-cytoplasmic male-sterility. *Proc. Roy. Soc. London B* 251: 253–61.

Mayer WE, Herrmann M, and Sommer RJ. 2007. Phylogeny of the nematode genus *Pristionchus* and implications for biodiversity, biogeography and the evolution of hermaphroditism. *BMC Evol. Biol.* 7: 104–116.

Maynard Smith J. 1976. Evolution and the theory of games. *Amer. Sci.* 64: 41–45.

Maynard Smith J. 1978. *The Evolution of Sex*. Cambridge: Cambridge UP.

Maynard Smith J. 1982. *Evolution and the Theory of Games*. New York: Cambridge UP.

Maynard Smith J. 1998. *Evolutionary Genetics*. 2nd ed. Oxford: Oxford UP.

Mazer SJ, Delasalle VA, and Paz H. 2007. Evolution of mating system and the genetic covariance between male and female investment in *Clarkia* (Onagraceae): Selfing opposes the evolution of trade-offs. *Evolution* 61: 83–98.

McMullen CK. 1987. Breeding systems of selected Galapagos Islands angiosperms. *Amer. J. Bot.* 74: 1694–1705.

McMullen CK. 1990. Reproductive biology of Galapagos Islands angiosperms. *Monogr. Syst. Bot.* 32: 35–45.

Metzlaff, M., Borner, T., and Hagemann, R. 1981. Variations of chloroplast DNAs in the genus *Pelargonium* and their biparental inheritance. *Theoret. Appl. Genet.* 60: 37–41.

Micale V, Maricchiolo G, and Genovese L. 2002. The reproductive biology of blackspot sea bream *Pagellus bogaraveo* in captivity. I. Gonadal development, maturation, and hermaphroditism. *J. Appl. Ichthy.* 18: 172–76.

Michiels NK. 1998. Mating conflicts and sperm competition in simultaneous hermaphrodites. In T. R. Birkhead and A. P. Møller, eds., *Sperm Competition and Sexual Selection*, 219–54. London: Academic Press.

Michiels NK, and Bakovski B. 2000. Sperm trading in a hermaphroditic flatworm: Reluctant fathers and sexy mothers. *Anim. Behav.* 59: 319–25.

Michiels NK, Beukeboom LW, Greeff JM, and Pemberton AJ. 1999. Individual control over reproduction: An underestimated element in the maintenance of sex? *J. Evol. Biol.* 12: 1036–1039.

Michiels NK, and Koene JM. 2006. Sexual selection favours harmful mating in hermaphrodites more than in gonochorists. *Integr. Comp. Biol.* 46: 473–80.

Michiels NK, and Newman LJ. 1998. Sex and violence in hermaphrodites. *Nature* 391: 647.

Michiels NK, Raven-Yoo-Heufes A, and Brockmann K. 2003. Sperm trading and sex roles in the hermaphroditic opisthobranch sea slug *Navanax inermis*: Eager females or opportunistic males? *Biol. J. Linnean Soc.* 78: 105–116.

Milinski M. 2006. Fitness consequences of selfing and outcrossing in the cestode *Schistocephalus solidus*. *Integr. Comp. Biol.* 46: 373–80.

Miller JS, Levin RA, and Feliciano NM. 2008. A tale of two continents: Baker's rule and the maintenance of self-incompatibility in *Lycium* (Solanaceae). *Evolution* 62: 1052–1065.

Miller JS, and Venable DL. 2000. Polyploidy and the evolution of gender dimorphism in plants. *Science* 289: 2335–38.

Miura S, Komatsu T, Higa M, Bhandari RK, Nakamura S, and Nakamura M. 2003. Gonadal sex differentiation in protandrous anemone fish, *Amphiprion clarkii*. *Fish Physiol. Biochem.* 28: 165–66.

Moe MA. 1969. Biology of the red grouper *Epinephelus morio* (Valenciennes) from the eastern Gulf of Mexico. *Florida Dept. Natur. Res. Lab, Prof. Papers* 10: 1–95.

Moiseeva EB, Sachs O, Zak T, and Funkenstein B. 2001. Protandrous hermaphroditism in Australian silver perch, *Bidyanus bidyanus* (Mitchell 1836). *Israeli J. Aquacul.* 53: 57–68.

Morgan MT, and Conner JK. 2001. Using genetic markers to directly estimate male selection gradients. *Evolution* 55: 272–81.

Moyer JT, and Nakazono A. 1978. Population structure, reproductive behavior, and protogynous hermaphroditism in the angelfish *Centropyge interruptus*. *Japan. J. Ichthyol.* 25: 101–106.

Muenchow GE. 1987. Is dioecy associated with fleshy fruit? *Am. J. Bot.* 74: 287–93.

Munday PL. 2002. Bi-directional sex change: Testing the growth-rate advantage model. *Behav. Ecol. Sociobiol.* 52: 247–54.

Munday PL, Buston PM, and Warner RR. 2006. Diversity and flexibility of sex-change strategies in animals. *Trends Ecol. Evol.* 21: 89–95.

Munday PL, Caley MJ, and Jones GP. 1998. Bi-directional sex change in a coral-dwelling goby. *Behav. Ecol. Sociobiol.* 43: 371–77.

Munday PL, and Molony BW. 2002. Energetic costs of protogyny versus protandry in the bidirectional sex changing fish, *Gobiodon histrio. Marine Biol.* 141: 1011–1017.

Muñoz RC, and Warner RR. 2003. A new version of the size-advantage hypothesis for sex change: Incorporating sperm competition and size-fecundity skew. *Amer. Natur.* 161: 749–61.

Muñoz RC, and Warner RR. 2004. Testing a new version of the size-advantage hypothesis for sex change: Sperm competition and size skew effects in the bucktooth parrotfish, *Sparisoma radians. Behav. Ecol.* 15: 129–36.

Murphy WJ, Thomerson JE, and Collier GE. 1999. Phylogeny of the neotropical killifish family Rivulidae (Cyprinodontiformes, Aplocheiloidei) inferred from mitochondrial DNA sequences. *Mol. Phylogen. Evol.* 13: 289–301.

Nagylaki T. 1976. A model for the evolution of self-fertilization and vegetative reproduction. *J. Theoret. Biol.* 58: 55–58.

Nakashima Y, Kuwamura T, and Yogo Y. 1995. Why be a both-ways sex changer. *Ethology* 101: 301–307.

Nakashima Y, Kuwamura T, and Yogo Y. 1996. Both-ways sex change in monogamous coral gobies, *Gobiodon* spp. *Environ. Biol. Fish.* 46: 281–88.

Nash JP, Kime DE, Van der Van LTM, Wester PW, Brion F, Maack G, Stahlschmidt-Allner P, and Tyler CR. 2004. Long-term exposure to environmental concentrations of the pharmaceutical ethynylestradiol causes reproductive failure in fish. *Environ. Health Perspect.* 112: 1725–33.

Nayak S, Goree J, and Schedl T. 2005. *fog-2* and the evolution of self-fertile hermaphroditism in *Caenorhabditis. PLos Biol.* 3: 57–71.

Neigel JE, and Avise JC. 1983. Clonal diversity and population structure in a reef-building coral, *Acropora cervicornis*: Self-recognition analysis and demographic interpretation. *Evolution* 37: 437–53.

Newbigin E, Anderson MA, and Clarke AE. 1993. Gametophytic self-incompatibility systems. *Plant Cell* 5: 1315–24.

Nigon V. 1949. Les modalites de la reproduction et le determinisme de sexe chez quelques Nematodes libres. *Ann. Sci. Nat. Zool. Ser.* 1: 1–132.

Nikolski GV. 1963. *The Ecology of Fishes.* London: Academic Press.

Nolan M, Jobling S, Brighty G, Sumpter JP, and Tyler CR. 2001. A histological description of intersexuality in the roach. *J. Fish Biol.* 58: 160–76.

Ohbayashi-Hodoki K, Ishihama F, and Shimada M. 2004. Body size-dependent gender role in a simultaneous hermaphrodite freshwater snail, *Physa acuta. Behav. Ecol.* 15: 976–81.

Oldfield RG. 2005. Genetic, abiotic and social influences on sex differentiation in cichlid fishes and the evolution of sequential hermaphroditism. *Fish and Fisheries* 6: 93–110.

Osborne MJ. 2004. *An Introduction to Game Theory.* New York: Oxford UP.

Pannell JR. 1997. The maintenance of gynodioecy and androdioecy in a metapopulation. *Evolution* 51: 10–20.

Pannell JR. 2002. The evolution and maintenance of androdioecy. *Annu. Rev. Ecol. Syst.* 33: 397–425.

Pannell JR. 2007. Variation in sex ratios and sex allocation in androdioecious *Mercurialis annua. J. Ecol.* 85: 57–69.

Pannell JR. 2008. Consequences of inbreeding depression due to sex-linked loci for the maintenance of males and outcrossing in branchiopod crustaceans. *Genet. Res.* 90: 73–84.

Pannell JR. 2009. On the problems of a closed marriage: Celebrating Darwin 200. *Biol. Lett.* 5: 332–35.

Pannell JR, and Barrett SCH. 1998. Baker's law revisited: Reproductive assurance in a metapopulation. *Evolution* 52: 657–88.

Parker GA. 1970. Sperm competition and its evolutionary consequences in the insects. *Biol. Rev.* 45: 525–67.

Parker GA. 1984. Sperm competition and the evolution of male mating strategies. In R. L. Smith, ed., *Sperm Competition and the Evolution of Animal Mating Systems,* 1–60. New York: Academic Press.

Parker GA. 2006. Sexual conflict over mating and fertilization: An overview. *Phil. Trans. Royal Soc. Lond. B* 361: 235–59.

Parker GA, Baker RR, and Smith VGF. 1972. The origin and evolution of gamete dimorphism and the male-female phenomenon. *J. Theoret. Biol.* 36: 529–53.

Parker GA, and Simmons LW. 1996. Parental investment and the control of sexual selection: Predicting the direction of sexual competition. *Proc. Royal Soc. London B* 263: 315–21.

Parker PG, and Tang-Martinez Z. 2005. Bateman gradients in field and laboratory studies: A cautionary tale. *Integr. Comp. Biol.* 45: 895–902.

Patino R, and 7 others. 1996. Sex differentiation of channel catfish gonads: Normal development and effects of temperature. *J. Exptl. Zool.* 276: 209–218.

Pauly D. 2004. *Darwin's Fishes.* Cambridge: Cambridge UP.

Payne RB. 1979. Sexual selection and intersexual differences in variance of breeding success. *Amer. Natur.* 114: 447–52.

Pearse DE, and Avise JC. 2001. Turtle mating systems: Behavior, sperm storage, and genetic paternity. *J. Heredity* 92: 206–211.

Pearse DE, Janzen FJ, and Avise JC. 2001. Genetic markers substantiate long-term storage and utilization of sperm by female painted turtles. *Heredity* 86: 378–84.

Penman DJ, and Piferrer F. 2008. Fish gonadogenesis. Part I: Genetic and environmental mechanisms of sex determination. *Rev. Fish Sci.* 16(S1): 16-34.

Petersen CW. 1987. Reproductive behaviour and gender allocation in *Serranus fasciatus*, a hermaphroditic reef fish. *Anim. Behav.* 35: 1601–1614.

Petersen CW. 1990. Variation in reproductive success and gonadal allocation in the simultaneous hermaphrodite *Serranus fasciatus*. *Oecologia* 83: 62–67.

Petersen CW. 1995. Reproductive behavior, egg trading, and correlates of male mating success in the simultaneous hermaphrodite, *Serranus tabacarius*. *Environ. Biol. Fish.* 43: 351–61.

Petersen CW, and Fischer EA. 1986. Mating system of the hermaphroditic coral-reef fish, *Serranus baldwini*. *Behav. Ecol. Sociobiol.* 19: 171–78.

Pianka ER. 1988. *Evolutionary Ecology.* 4th ed. New York: Harper and Row.

Policansky D. 1981. Sex choice and the size advantage model in jack-in-the-pulpit (*Arisaema triphyllum*). *Proc. Natl. Acad. Sci. USA* 78: 1306–1308.

Policansky D. 1982. Sex change in plants and animals. *Annu. Rev. Ecol. Syst.* 13: 471–95.

Pongratz N, and Michiels NK. 2003. High multiple paternity and low last-male sperm precedence in a hermaphroditic planarian flatworm: Consequences for reciprocity patterns. *Mol. Ecol.* 12: 1425–33.

Poole CF, and Geimball PC. 1939. Inheritance of new sex forms of *Cucumis melo* L. *J. Heredity* 30: 21–25.

Potrzebowski L, Vinckenbosch N, Marques AC, Chalmel F, Jegou B, and Kaessmann H. 2008. Chromosomal gene movements reflect the recent origin and biology of therian sex chromosomes. *PloS Biol.* 6: 709–716.

Power AJ, and Keegan BF. 2001. The significance of imposex levels and TBT contamination in the red whelk, *Neptunia antigua* (L.) from the offshore Irish Sea. *Marine Pollution Bull.* 42: 761–72.

Prahlad V, Pilgrim D, and Goodwin EB. 2003. Roles for mating and environment in *C. elegans* sex determination. *Science* 302: 1046–1049.

Premoli MC, and Sella G. 1995. Sex economy in benthic polychaetes. *Ethol. Ecol. Evol.* 7: 27–48.

Price SC, and Jain SK. 1981. Are inbreeders better colonizers? *Oecologia* 49: 283–86.

Primack RB, and Lloyd DG. 1980. Sexual strategies in plants IV. The distribution of gender in two monomorphic shrub populations. *New Zeal. J. Bot.* 18: 109–114.

Pryke SR, and Griffith SC. 2009. Genetic incompatibility drives sex allocation and maternal investment in a polymorphic finch. *Science* 232: 1605–1607.

Puurtinen M, and Kaitala V. 2002. Mate-search efficiency can determine the evolution of separate sexes and the stability of hermaphroditism in animals. *Amer. Natur.* 160: 645–60.

Quagio-Grassiotto I, Carvalho ED. 1999. The ultrastructure of *Sorubim lima* (Teleostei, Siluriformes, Pimelodidae) spermatogenesis: premeiotic and meiotic periods. *Tissue and Cell* 31: 561–67.

Raimondi PT, and Martin JE. 1991. Evidence that mating group size affects allocation of reproductive resources in a simultaneous hermaphrodite. *Amer. Natur.* 138: 1206–1217.

Ralls K, and Ballou J. 1983. Extinction: Lessons from zoos. In C. M. Shonewald-Cox, S. M. Chambers, B. MacBryde, and L. Thomas, eds., *Genetics and Conservation*, 164–84. Menlo Park, Calif.: Benjamin/Cummings.

Rambuda TD, and Johnson SD. 2004. Breeding systems of invasive alien plants in South Africa: Does Baker's rule apply? *Diversity Distrib.* 10: 409–416.

Raven PH. 1979. A survey of reproductive biology in Onagraceae. *N. Zeal. J. Bot.* 17: 575–93.

Renner SS. 1998. Phylogenetic affinities of Monimiaceae based on cpDNA gene and spacer sequences. *Persp. Plant Ecol., Evol. Syst.* 1: 61–77.

Renner SS, and Feil JP. 1993. Pollinations of tropical dioecious angiosperms. *Amer. J. Bot.* 80: 1100–1107.

Renner SS, and Ricklefs RE. 1995. Dioecy and its correlates in flowering plants. *Amer. J. Bot.* 82: 596–606.

Renner SS, and Won H. 2001. Repeated evolution of monoecy in Siparunaceae (Laurales). *Syst. Biol.* 50: 700–712.

Ribeiro de Oliveira R, Souza IL, and Venere PC. 2008. Karyotype description of three species of Loricariidae (Siluriformes) and occurrence of the ZZ/ZW sexual system in *Hemiancistrus spilomma*. *Neotropical Ichthyol.* 4: 93–97.

Riddle LD, Blumenthal T, Meyer BJ, and Priess JR, eds. 1997. *C. elegans II.* Cold Spring Harbor, N.Y.: Cold Spring Harbor Laboratory Press.

Rieseberg LH, Hanson MA, and Philbrick CT. 1992. Androdieocy is derived from dioecy in Datiscaceae: Evidence from restriction site mapping of PCR-amplified chloroplast DNA fragments. *Syst. Bot.* 17: 324–36.

Ritland C, and Ritland K. 1989. Variation of sex allocation among eight taxa of the *Mimulus guttatus* species complex. *Amer. J. Bot.* 76: 1731–39.

Robertson DR. 1972. Social control of sex reversal in coral reef fish. *Science* 177: 1007–1009.

Robertson DR, and Justines G. 1982. Protogynous hermaphroditism and gonochorism in four Caribbean reef gobies. *Environ. Biol. Fish.* 7: 137–42.

Robertson DR, Reinboth R, and Bruce RW. 1982. Gonochorism, protogynous sex-change and spawning in three sparisomatinine parrotfishes from the western Indian Ocean. *Bull. Mar. Sci.* 32: 868–79.

Robertson DR, and Warner RR. 1978. Sexual patterns in the labroid fishes of the western Caribbean, 2: The parrotfishes (Scaridae). *Smithson. Contrib. Zool.* 255: 1–26.

Rodgers EW, Earley RL, and Grober MS. 2007. Social status determines sexual phenotype in the bi-directional sex changing blue banded goby *Lythrypnus dalli*. *J. Fish Biol.* 70: 1660–68.

Rogers D, and Chase R. 2001. Dart receipt promotes sperm storage in the garden snail *Helix aspersa*. *Behav. Ecol. Sociobiol.* 50: 122–27.

Rogers D, and Chase R. 2002. Determinants of paternity in the garden snail *Helix aspersa*. *Behav. Ecol. Sociobiol.* 52: 289–95.

Rosas F, and Dominguez CA. 2008. Male sterililty, fitness gain curves and the evolution of gender specialization from distyly in *Erythroxylum havanense*. *J. Evol. Biol.* 22: 50–59.

Ross RM. 1990. The evolution of sex-change mechanisms in fishes. *Environ. Biol. Fish.* 29: 81–93.

Ross RM, Losey GS, and Diamond M. 1983. Sex change in a coral-reef fish: Dependence of stimulation and inhibition on relative size. *Science* 221: 574–75.

Rubin DA. 1985. Effects of pH on sex ratios in cichlids and a poeciliid (Teleostei). *Copeia* 1985: 233–35.

Sadovy Y, and Liu M. 2008. Functional hermaphroditism in teleosts. *Fish and Fisheries* 9: 1–43.

Sadovy Y, and Shapiro DY. 1987. Criteria for the diagnosis of hermaphroditism in fishes. *Copeia* 1987: 136–56.

Sakai AK. 2001. Thrips pollination of androdioecious *Castilla elastica* (Moraceae) in a seasonal tropical forest. *Amer. J. Bot.* 88: 1527–34.

Sakai AK, Wagner WL, Ferguson DM, and Herbst DR. 1995a. Biogeographical and ecological correlates of dioecy in the Hawaiian flora. *Ecology* 76: 2530–43.

Sakai AK, Wagner WL, Ferguson DM, and Herbst DR. 1995b. Origins of dioecy in the Hawaiian flora. *Ecology* 76: 2517–29.

Sakai AK, and Weller SG. 1999. Gender and sexual dimorphism in flowering plants: A review of terminology, biogeographic patterns, ecological correlates, and phylogenetic approaches. In E. Geber, T. E. Dawson, and L. F. Delph, eds., *Gender and Sexual Dimorphism in Flowering Plants*, 1–31. Berlin: Springer-Verlag.

Sakai AK, Weller SG, Wagner WL, Nepokroeff M, and Culley TM. 2006. Adaptive radiation and evolution of breeding systems in *Schiedea* (Caryophyllaceae), an endemic Hawaiian genus. *Ann. Missouri Bot. Gard.* 93: 49–63.

Sakakura Y, and Noakes DLG. 2000. Age, growth, and sexual development in the self-fertilizing hermaphroditic fish *Rivulus marmoratus*. *Environ. Biol. Fishes* 59: 309–317.

Sassaman C, and Weeks SC. 1993. The genetic mechanism of sex determination in the conchostracan shrimp *Eulimnadia texana*. *Amer. Natur.* 141: 314–28.

Sato A, and 7 others. 2002. Persistence of *Mhc* heterozygosity in homozygous clonal killifish, *Rivulus marmoratus*: Implications for the origin of hermaphroditism. *Genetics* 162: 1791–1803.

Savage AE, and Miller JS. 2006. Gametophytic self-incompatibility in *Lycium parishii* (Solanaceae): Allelic diversity, genealogical structure, and patterns of molecular evolution at the S-RNase locus. *Heredity* 96: 434–44.

Schärer L. 2009. Tests of sex allocation theory in simultaneously hermaphroditic animals. *Evolution* 63: 1377–1405.

Schärer L, and Janicke T. 2009. Sex allocation and sexual conflict in simultaneously hermaphroditic animals. *Biol. Lett. Evol. Biol.* 5: 705–708.

Schärer L, Karlsson LM, Christen M, and Wedekind C. 2001. Size-dependent sex allocation in a simultaneous hermaphrodite parasite. *J. Evol. Biol.* 14: 55–67.

Schärer L, and Ladurner P. 2003. Phenotypically plastic adjustment of sex allocation in a simultaneous hermaphrodite. *Proc. Roy. Soc. London B* 270: 935–41.

Schärer L, Ladurner P, and Rieger RM. 2004. Bigger testes do work more: Experimental evidence that testis size reflects testicular cell proliferation activity in a marine invertebrate, the free-living flatworm *Macrostomum* sp. *Behav. Ecol. Sociobiol.* 56: 420–25.

Schärer L, Sandner P, and Michiels NK. 2005. Trade-off between male and female allocation in the simultaneously hermaphroditic flatworm *Macrostomum* sp. *J. Evol. Biol.* 18: 396–404.

Schärer L, and Wedekind C. 2001. Social situation, sperm competition, and sex allocation in a simultaneous hermaphrodite parasite, the cestode *Schistocephalus solidus. J. Evol. Biol.* 14: 942–53.

Schemske DW, and Lande R. 1985. The evolution of self-fertilization and inbreeding depression in plants. II. Empirical observations. *Evolution* 39: 41–52.

Schierup MH, and Christiansen FB. 1996. Inbreeding depression and outbreeding depression in plants. *Heredity* 77: 461–68.

Schierup MH, Vekemans X, and Christiansen FB. 1998. Allelic genealogies in sporophytic self-incompatibility systems in plants. *Genetics* 150: 1187–98.

Schjørring S. 2004. Delayed selfing in relation to the availability of a mating partner in the cestode *Schistocephalus solidus. Evolution* 58: 2591–96.

Schnable PS, and Wise RP. 1998. The molecular basis of cytoplasmic male sterility and fertility restoration. *Trends Plant Sci.* 3: 175–80.

Schrag SJ, and Read AF. 1996. Loss of male outcrossing ability in simultaneous hermaphrodites: phylogenetic analyses of pulmonate snails. *J. Zool.* 238: 287–99.

Schueller SK. 2004. Self-pollination in island and mainland populations of the introduced hummingbird-pollinated plant, *Nicotiana glauca* (Solanaceae). *Amer. J. Bot.* 91: 672–81.

Schultz, R. J. 1993. Genetic regulation of temperature-mediated sex ratios in the livebearing fish *Poeciliopsis lucida. Copeia* 1993: 1148–51.

Scofield VL, Schlumpberger JM, West LA, and Weissman IL. 1982. Protochordate allorecognition is controlled by a MHC-like gene system. *Nature* 295: 499–502.

Seger J, and Eckhart VM. 1996. Evolution of sexual systems and sex allocation in plants when growth and reproduction overlap. *Proc. Royal Soc. London B* 263: 833–41.

Selander RK. 1975. Stochastic factors in the genetic structure of populations. In G. F. Estabrook, ed., *Proceedings of the Eight International Conference on Numerical Taxonomy*, 284–332. San Francisco: Freeman.

Selander RK, and Hudson RO. 1976. Animal population structure under close inbreeding: The land snail *Rumina* in southern France. *Amer. Natur.* 110: 695–718.

Selander RK, and Kaufman DW. 1973. Self-fertilization and genetic population structure in a colonizing land snail. *Proc. Natl. Acad. Sci. USA* 70: 1186–90.

Selander RK, and Kaufman DW. 1975. Genetic population structure and breeding systems. *Isozymes IV:* 27–48.

Selander RK, Kaufman DW, and Ralin RS. 1974. Self-fertilization in the terrestrial snail *Rumina decollata. Veliger* 16: 265–70.

Selander RK, and Ochman H. 1983. The genetic structure of populations as illustrated by mollusks. *Isozymes* 10: 93–123.

Sella G. 1985. Reciprocal egg trading and brood care in a hermaphroditic polychaete worm. *Anim. Behav.* 33: 938–44.

Sella G. 1988. Reciprocation, reproductive success, and safeguards against cheating in a hermaphroditic polychaete worm, *Ophryotrocha diadema* Akesson, 1976. *Biol. Bull.* 175: 212–17.

Sella G, and Lorenzi MC. 2003. Increased sperm allocation delays body growth in a protandrous simultaneous hermaphrodite. *Biol. J. Linnean Soc.* 78: 149–54.

Sella G, Premoli MC, and Turri F. 1997. Egg trading in the simultaneously hermaphroditic polychaete worm *Ophryotrocha gracilis* (Huth.). *Behav. Ecol.* 8: 83–86.

Sella G, and Ramella L. 1999. Sexual conflict and mating systems in the dorvilleid genus *Ophyrotrocha* and the dinophilid genus *Dinophilus. Hydrobiologia* 402: 203–213.

Shapiro DY. 1979. Social behavior, group structure, and the control of sex reversal in hermaphroditic fish. *Adv. Stud. Behav.* 10: 43–102.

Shapiro DY. 1987. Differentiation and evolution of sex change in fishes: A coral reef fish's social environment can control its sex. *BioScience* 37: 490–97.

Shapiro DY. 1991. Intraspecific variability in social systems of coral reef fishes. In P. F. Sale, ed., *The Ecology of Fishes on Coral Reefs,* 331–55. San Diego: Acadmic Press.

Shapiro DY, and Lubbock R. 1980. Group sex ratio and sex reversal. *J. Theor. Biol.* 82: 411–26.

Shine R. 1999. Why is sex determined by nest temperature in many reptiles? *Trends Ecol. Evol.* 14: 186–89.

Shuster SM, and Wade MJ. 2003. *Mating Systems and Strategies.* Princeton: Princeton UP.

Skyrms B. 1996. *Evolution of the Social Contract.* Cambridge: Cambridge UP.

Smith CL. 1965. The patterns of sexuality and the classification of serranid fishes. *Amer. Museum Novit.* 2207: 1–20.

Smith CL. 1967. Contribution to a theory of hermaphroditism. *J. Theoret. Biol.* 17: 76–90.

Smith CL. 1975. The evolution of hermaphroditism in fishes. In R. Reinboth, ed., *Intersexuality in the Animal Kingdom*, 295–310. New York: Springer-Verlag.

Smith RL, ed. 1984. *Sperm Competition and the Evolution of Animal Mating Systems*. New York: Academic Press.

Snow AA, Spira TP, Simpson R, and Klips RA. 1996. The ecology of geitonogamous pollination. In D. G. Lloyd and S. C. H. Barrett, eds., *Floral Biology: Studies on Floral Evolution in Animal-pollinated Plants*, 191–216. New York: Chapman & Hall.

Snyder BF, and Gowaty PA. 2007. A reappraisal of Bateman's classic study of intrasexual selection. *Evolution* 61: 2457–68.

Solomon BP. 1986. Sexual allocation and andromonoecy: Resource investment in male and hermaphroditic flowers of *Solanum carolinense* (Solanaceae). *Amer. J. Bot.* 73: 1215–21.

Soltis PS, Soltis DE, Wolf PG, Nickrent DL, Chaw S-M, and Chapman RL. 1999. The phylogeny of land plants inferred from 18S rRNA sequences: Pushing the limits of rDNA signal? *Mol. Biol. Evol.* 16: 1774–84.

Stanton ML, Snow AA, and Handel SN. 1986. Floral evolution: Attractiveness to pollinators increases male fitness. *Science* 232: 1625–27.

Stebbins GL. 1970. Adaptive radiation in angiosperms. I. Pollination mechanisms. *Annu. Rev. Ecol. Syst.* 1: 307–326.

St. Mary CM. 1993. Novel sexual patterns in two simultaneously hermaphroditic gobies, *Lythrypnus dalli* and *Lythrypnus zebra*. *Copeia* 1993: 1062–1072.

St. Mary CM. 1994. Sex allocation in a simultaneous hermaphrodite, the blue-banded goby (*Lythrypnus dalli*): The effects of body size and behavioural gender and the consequences for reproduction. *Behav. Ecol.* 5: 304–313.

St. Mary CM. 1997. Sequential patterns of sex allocation in simultaneous hermaphrodites: Do we need models that specifically incorporate this complexity? *Amer. Natur.* 150: 73–97.

St. Mary CM. 1998. Characteristic gonad structure in the gobiid genus *Lythrypnus* with comparisons to other hermaphroditic gobies. *Copeia* 1998: 720–24.

Stone JL. 1995. Pollen donation patterns in a tropical distylous shrub (*Psychotria suerrensis* Rubiaceae). *Amer. J. Bot.* 82: 1390–98.

Strathmann RR, Strathmann MF, and Emson RH. 1984. Does limited brood capacity link adult body size, brooding, and simultaneous hermaphroditism? A test with the starfish *Asterina phylactica*. *Amer. Natur.* 123: 796–818.

Sun M. 1987. Genetics of gynodioecy in Hawaiian *Bidens* (Asteraceae). *Heredity* 59: 327–36.

Sun M, and Ritland K. 1998. Mating system of yellow starthistle (*Centaurea solstitialis*), a successful colonizer in North America. *Heredity* 80: 225–32.

Sunobe T, and Nakazono A. 1993. Sex change in both directions by alteration of social dominance in *Trimma okinawae* (Pisces: Gobiidae). *Ethology* 94: 339–45.

Swensen SW, Luthi JN, and Rieseberg LH. 1998. Datiscaceae revisited: Monophyly and the sequence of breeding system evolution. *Syst. Bot.* 23: 157–69.

Taborsky M. 2008. Alternative reproductive tactics in fish. In R. F. Oliveira, M. Taborsky, and H. J. Brockmann, eds., *Alternative Reproductive Tactics*, 251–99. Cambridge: Cambridge UP.

Tan GN, Govedich FR, and Burd M. 2004. Social group size, potential sperm competition and reproductive investment in a hermaphroditic leech, *Helobdella papillornata* (Euhirudinea: Glossiphoniidae). *J. Evol. Biol.* 17: 575–80.

Tang-Martinez A, and Ryder TB. 2005. The problem with paradigms: Bateman's worldview as a case study. *Integr. Comp. Biol.* 45: 821–30.

Tatarenkov A, and Avise JC. 2007. Rapid concerted evolution in animal mitochondrial DNA. *Proc. Roy. Soc. Lond. B* 274: 1795–98.

Tatarenkov A, Gao H, Mackiewicz M, Taylor DS, Turner BJ, and Avise JC. 2007. Strong population structure despite evidence of recent migration in a selfing hermaphroditic vertebrate, the mangrove killifish (*Kryptolebias marmoratus*) *Mol. Ecol.* 16: 2701–2711.

Tatarenkov A, Lima SMQ, Taylor DS, and Avise JC. 2009. Long-term retention of self-fertilization in a fish clade. *Proc. Natl. Acad. Sci. USA* 106: 14456–459.

Taylor DS. 1990. Adaptive specializations of the cyprinodont fish, *Rivulus marmoratus. Florida Sci.* 53: 239–48.

Taylor DS. 2000. Biology and ecology of *Rivulus marmoratus*: New insights and a review. *Florida Sci.* 63: 242–55.

Taylor PD. 1981. Intra-sex and inter-sex sibling interactions as sex ratio determinants. *Nature* 291: 64–66.

Taylor RG, Whittington JA, Grier HJ, and Crabtree RE. 2000. Age, growth, maturation, and protandric sex reversal in common snook, *Centropomus undecimalis*, from the east and west coasts of South Florida. *Fish. Bull.* 98: 612–24.

Taylor DS, Fisher MT, and Turner BJ. 2001. Homozygosity and heterozygosity in three populations of *Rivulus marmoratus. Environ. Biol. Fish.* 61: 455–59.

Thompson JD, and Brunet J. 1990. Hypotheses for the evolution of dioecy in seed plants. *Trends Ecol. Evol.* 5: 11–16.

Tomiyama K. 1996. Mate-choice criteria in a protandrous simultaneously hermaphroditic land snail *Achatina fulica* (Férussac) (Stylommatophora: Achatinidae). *J. Moll. Stud.* 62: 101–111.

Tomlinson J. 1966. The advantages of hermaphroditism and parthenogenesis. *J. Theoret. Biol.* 11: 54–58.

Trivers RL. 1972. Parental investment and sexual selection. In B. Campbell, ed., *Sexual Selection and the Descent of Man*, 136–79. Chicago: Aldine.

Trouvé S, Jourdane J, Renaul F, Durand P, and Morand S. 1999. Adaptive sex allocation in a simultaneous hermaphrodite. *Evolution* 53: 1599–1604.

Tsitrone A, Jarne P, and David P. 2003. Delayed selfing and resource allocations in relation to mate availability in the freshwater snail *Physa acuta. Amer. Natur.* 162: 474–88.

Turner BJ, Elder JF Jr., Laughlin TF, and Davis WP. 1990. Genetic variation in clonal vertebrates detected by simple-sequence DNA fingerprinting. *Proc. Natl. Acad. Sci. USA* 87: 5653–57.

Turner BJ, Davis WP, and Taylor DS. 1992a. Abundant males in populations of a selfing hermaphroditic fish—*Rivulus marmoratus*—from some Belize Cays. *J. Fish Biol.* 40: 307–310.

Turner BJ, Elder JF Jr., Laughlin TF, Davis WP, and Taylor DS. 1992b. Extreme clonal diversity and divergence in popuations of the self-fertilizing killifish, *Kryptolebias marmoratus. Proc. Natl. Acad. Sci. USA* 89: 10643–10647.

Turner BJ, Fisher MT, Taylor DS, Davis WP, and Jarrett BL. 2006. Evolution of "maleness" and outcrossing in a population of the self-fertilizing killifish, *Kryptolebias marmoratus. Evol. Ecol. Res.* 8: 1475–86.

Ueno K, and Takai A. 2008. Multiple sex chromosome system of X1X1X2X2/ X1X2Y type in lutjanid fish, *Lutjanus quinquelineatus* (Perciformes) *Genetica* 132: 35–41.

Uller T, Pen I, Wapstra E, Beukeboom LW, and Komdeur J. 2007. The evolution of sex ratios and sex-determining systems. *Trends Ecol. Evol.* 22: 292–97.

Uyenoyama M, Holsinger K, and Waller D. 1993. Ecological and genetic factors directing the evolution of self-fertilization. *Oxford Surv. Evol. Biol.* 9: 327–81.

Valenzuela N, Adams DC, and Janzen, FJ. 2003. Pattern does not equal process: Exactly when is sex environmentally determined? *Amer. Natur.* 161: 676–83.

Vamosi JC, Otto SP, and Barrett SCH. 2003. Phylogenetic analysis of the ecological correlates of dioecy in angiosperms. *J. Evol. Biol.* 16: 1006–1018.

Vamosi JC, Zhang Y, and Wilson WG (2007) Animal dispersal dynamics promoting dioecy over hermaphroditism. *Amer. Natur.* 170: 485–91.

Van Kleunen M, and Ritland K. 2004. Predicting evolution of floral traits associated with mating system in a natural population. *J. Evol. Biol.* 17: 1389–99.

Vandeputte M, Dupont-Nivet M, Chavanne H, and Chatain B. 2007. A polygenic hypothesis for sex determination in the European sea bass—*Dicentrarchus labrax. Genetics* 176: 1049–1057.

Vassiliadis C, Saumitou-Laprade P, Lepart J, and Viard F. 2002. High male reproductive success of hermaphrodites in the androdioecious *Phillyrea angustifolia. Evolution* 56: 1362–73.

Veyrunes, F., and 14 others. 2008. Bird-like sex chromosomes of platypus imply recent origin of mammal sex chromosomes. *Genome Res.* 18: 965–73.

Viard F, Doums C, and Jarne P. 1997. Selfing, sexual polymorphism and microsatellites in the hermaphroditic freshwater snail *Bulinus truncatus. Proc. Roy. Soc. London B* 264: 39–44.

Vicari MR, Artoni RF, Moreira-Filgo O, and Bertollo LAC. 2008. Diversification of a ZZ/ZW sex chromosome system in Characidium fish (Crenuchidae, Characiformes). *Genetica* 134: 311–17.

Vincent A, Ahnesjö I, Berglund A, and Rosenqvist G. 1992. Pipefishes and seahorses: Are they all sex role reversed? *Trends Ecol. Evol.* 7: 237–41.

Vine E, Shears J, van Aerle R, Tyler CR, and Sumpter JP. 2005. Endocrine (sexual) disruption is not a prominent feature of the pike (*Esox lucius*), a top predator, living in English waters. *Environ. Toxicol. Chem.* 24: 1436–43.

Vitturi R, and Catalano E. 1988. A male XO sex-determining mechanism in *Theodoxus meridionalis* (Neritidae) (Prosobranchia, Archaeogastropoda). *Cytologia* 53: 131–38.

Vitturi R, Catalano E, Macaluso M, and Zava B. 1988. The karyology of *Littorina neritoides* (Linnaeus, 1758) (Mollusca, Prosobranchia). *Malacologia* 29: 310–24.

Vitturi R, Colomba MS, Caputo V, and Pandolfo A. 1998. XY chromosome sex systems in the neogastropods *Fasciolaria lignaria* and *Pisania striata* (Mollusca: Prosobranchia). *J. Heredity* 89: 538–43.

Vitturi R, Libertini A, Panozzo M, and Mezzapelle G. 1995. Karyotype analysis and genome size in three Mediterranean species of periwinkles (Prosobranchia: Mesogastropoda). *Malacologia* 37: 123–32.

Vogler DW, and Kalisz S. 2001. Sex among the flowers: The distribution of plant mating systems. *Evolution* 55: 202–204.

Vreys C, and Michiels NK. 1997. Flatworms flatten to size up each other. *Proc. Roy. Soc. London B* 264: 1559–64.

Vreys C, and Michiels NK. 1998. Sperm trading by volume in a hermaphroditic flatworm with mutual penis intromission. *Anim. Behav.* 56: 777–85.

Vrijenhoek RC. 1985. Homozygosity and interstrain variation in the self-fertilizing hermaphroditic fish, *Rivulus marmoratus*. *J. Heredity* 76: 1475–86.

Wade MJ, and Arnold SJ. 1980. The intensity of sexual selection in relation to male sexual behavior, female choice, and sperm precedence. *Anim. Behav.* 28: 446–61.

Wade MJ, and Shuster SM. 2004. Sexual selection: Harem size and the variance in male reproductive success. *Amer. Natur.* 164: E83–89.

Warner RR. 1975. The adaptive significance of sequential hermaphroditism in animals. *Amer. Natur.* 109: 61–82.

Warner RR. 1982. Mating systems, sex change, and sexual demography in the rainbow wrasse, *Thalassoma lucasanum*. *Copeia* 3: 653–61.

Warner RR. 1984. Mating behavior and hermaphroditism in coral reef fishes. *Amer. Sci.* 72: 128–36.

Warner RR. 1988. Sex change and the size-advantage model. *Trends Ecol. Evol.* 3: 133–36.

Warner RR. 1991. The use of phenotypic plasticity in coral reef fishes as tests of theory in evolutionary ecology. In P. F. Sale, ed., *The Ecology of Fishes on Coral Reefs*, 387–98. San Diego: Academic Press.

Warner RR, Fitch DL, and Standish JD. 1996. Social control of sex change in the shelf limpet, *Crepidula norrisiarum*: size-specific responses to local group composition. *J. Exptl. Marine Biol. Ecol.* 204: 155–67.

Warner RR, and Hoffman SG. 1980. Local population size as a determinant of mating system and sexual composition in two tropical marine fishes (*Thalassoma* sp.). *Evolution* 34: 508–518.

Warner RR, and Robertson DR. 1978. Sexual patterns in the labroid fishes of the western Caribbean, 1: The wrasses (Labridae). *Smithson. Contrib. Zool.* 254: 1–27.

Warner RR, Robertson DR, and Leigh EG. 1975. Sex change and sexual selection. *Science* 190: 633–38.

Warner RR, and Swearer SE. 1991. Social control of sex change in the bluehead wrasse, *Thalassoma bifasciatum* (Pisces: Labridae). *Biol. Bull.* 181: 199–204.

Waser NM. 1993. Population structure, optimal outbreeding, and assortative mating in angiosperms. In N. W. Thornhill, ed., *The Natural History of Inbreeding and Outbreeding*, 173–99. Chicago: U of Chicago P.

Waser NM, and Price MV. 1994. Crossing-distance effects in *Delphinium nelsonii*: Outbreeding and inbreeding depression in progeny fitness. *Evolution* 48: 842–52.

Waters PD, Delbridge ML, Deakin JE, El-Mogharbel N, Kirby PJ, Carvalho-Silva DR, and Graves JM. 2005. Autosomal location of genes from the conserved mammalian X in the platypus (*Ornithorhynchus anatinus*): Implications for mammalian sex chromosome evolution. *Chromosome Res.* 13: 401–410.

Webb CJ. 1999. Empirical studies: Evolution and maintenance of dimorphic breeding systems. In M. A. Geber, T. E. Dawson, and L. F. Delph, eds., *Gender and Sexual Dimorphism in Flowering Plants*, 61–95. Berlin: Springer-Verlag.

Webb CJ, and Kelly D. 1993. The reproductive biology of the New Zealand flora. *Trends Ecol. Evol.* 8: 442–47.

Webb CJ, and Lloyd DG. 1986. The avoidance of interference between the presentation of pollen and stigmas in angiosperms. II. Herkogamy. *New Zealand J. Bot.* 24: 163–74.

Webster JP, and Gower CM. 2006. Mate choice, frequency dependence and the maintenance of resistance to parasitism in a simultaneous hermaphrodite. *Integr. Comp. Biol.* 46: 407–418.

Weeks SC, Benvenuto C, and Reed SK. 2006. When males and hermaphrodites coexist: A review of androdioecy in animals. *Integr. Comp. Biol.* 46: 449–64.

Weeks SC, Chapman EG, Rogers DC, Senyo DM, and Hoeh WR. 2009. Evolutionary transitions among dieocy, androdioecy and hermaphroditism in limnadiid clam shrimp (Branchiopod: Spinicaudata) *J. Evol. Biol.* 22: 1781–99.

Weeks SC, Posgai RT, Cesari M, and Scanabissi F. 2005a. Androdioecy inferred in the clam shrimp *Eulimnadia agassizii* (Spinicaudata: Limnadiidae). *J. Crustac. Biol.* 25: 323–28.

Weeks SC and 8 others. 2005b. Ancient androdioecy in the freshwater crustacean *Eulimnadia. Proc. Roy. Soc. Lond. B* 273: 725–34.

Weibel AC, Dowling TE, and Turner BJ. 1999. Evidence that an outcrossing population is a derived lineage in a hermaphroditic fish (*Rivulus marmoratus*). *Evolution* 53: 1217–25.

Weiblen GD, Oyama RK, and Donoghue MJ. 2000. Phylogenetic analysis of dioecy in monocotyledons. *Amer. Natur.* 155: 46–58.

Weir BS, and Cockerham CC. 1973. Mixed self and random mating at two loci. *Genet. Res.* 21: 247–62.

Weller SG, and Sakai AK. 1991. The genetic basis of male sterility in *Schiedea* (Caryophyllaceae), an endemic Hawaiian genus. *Heredity* 67: 265–73.

Weller SG, and Sakai AK. 1999. Using phylogenetic approaches for the analysis of plant breeding system evolution. *Annu. Rev. Ecol. Syst.* 30: 167–99.

Weller SG, Sakai AK, Wagner WL, and Herbst DR. 1990. Evolution of dioecy in *Schiedea* (Caryophyllaceae: Alsinoideae) in the Hawaiian Islands: Biogeographical and ecological factors. *Syst. Bot.* 15: 266–76.

Weller SG, Wagner WL, and Sakai AK. 1995. A phylogenetic analysis of *Schiedea* and *Alsinidendron* (Caryophyllaceae: Alsinoideae): Implications for the evolution of breeding systems. *Syst. Bot.* 20: 315–37.

Werren JH. 1998. Wolbachia and speciation. In D. Howard and S. Berlocher, eds., *Endless Forms: Species and Speciation*, 245–60. Oxford: Oxford UP.

West SA, Herre EA, and Sheldon BC. 2000. The benefit of allocating sex. *Science* 290: 288–90.

Westergaard M. 1958. The mechanism of sex determination in dioecious flowering plants. *Adv. Genet.* 9: 217–81.

Wethington AR, and Dillon RT. 1997. Selfing, outcrossing, and mixed mating in the freshwater snail *Physa heterostropha*: Lifetime fitness and inbreeding depression. *Invert. Biol.* 116: 192–99.

Williams GC. 1966. *Adaptation and Natural Selection*. Princeton: Princeton UP.

Williams GC. 1975. *Sex and Evolution*. Princeton: Princeton UP.

Williamson MH, and Fitter A. 1996. The characters of successful invaders. *Biol. Conserv.* 78: 163–70.

Willson MF. 1979. Sexual selection in plants: Perspective and overview. *Amer. Natur.* 113: 777–90.

Willson MF. 1983. *Plant Reproductive Ecology*. New York: Wiley.

Willson MF. 1990. Sexual selection in plants and animals. *Trends Ecol. Evol.* 5: 210–14.

Willson MF, and Burley N. 1983. Mate choice in plants: Tactics, mechanisms, and consequences. *Princeton Monographs in Population Biology* 19: 1–251.

Wolf DE, Satkoski JA, White K, and Rieseberg LH. 2001. Sex determination in the androdioecious plant *Datisca glomerata* and its dioecious sister species *D. cannabina*. *Genetics* 159: 1243–57.

Wolf DE, and Takebayashi N. 2004. Pollen limitation and the evolution of androdioecy from dioecy. *Amer. Natur.* 163: 122–37.

Wright S. 1931. Evolution in Mendelian populations. *Genetics* 16: 97–159.

Wright WG. 1989. Intraspecific density mediates sex-change in the territorial patellacean limpet *Lottia gagantea*. *Marine Biol.* 100: 353–64.

Wu G-C, Du J-L, Lee Y-H, Lee, M-F, and Chang C-F. 2005. Current status of genetic and endocrine factors in the sex change of protandrous black porgy, *Acanthopagrus schlegeli* (Teleostean). *Ann. N.Y. Acad. Sci.*1040: 206–214.

Wyatt R. 1988. Phylogenetic aspects of the evolution of self-pollination. In L. D. Gottlieb and S. K. Jain, eds., *Plant Evolutionary Biology*, 109–131. New York: Chapman & Hall.

Yampolsky E, and Yampolsky H. 1922. Distribution of sex forms in phanerogamic flora. *Bibl. Genet.* 3: 1–62.

Yano K, and Tanaka S. 1989. Hermaphroditism in the Lantern shark *Etmopterus unicolor* (Squalidae, Chondrichthyes). *Jap. J. Ichthyol.* 36: 338–45.

Young HH. 1937. *Genital Abnormalities, Hermaphroditism, and Related Adrenal Diseases.* New York: Williams & Wilkins.

Young PC, and Martin RB. 1985. Sex ratios and hermaphroditism in nemipterid fish from northern Australia. *J. Fish Biol.* 26: 273–87.

Yund PO. 1998. The effect of sperm competition on male gain curves in a colonial marine invertebrate. *Ecology* 79: 328–39.

Yund PO, Marcum Y, and Stewart-Savage J. 1997. Life-history variation in a colonial ascidian: Broad-sense heritabilities and tradeoffs in allocation to asexual growth and male and female reproduction. *Biol. Bull.* 192: 290–99.

Ziehe M, and Roberds JH. 1989. Inbreeding depression due to overdominance in partially self-fertilizing plant populations. *Genetics* 121: 861–68.

Zierold T, Hanfling B, and Gómez A. 2007. Recent evolution of alternative reproductive modes in the "living fossil" *Triops cancriformis*. *BMC Evol. Biol.* 7: 161–73.

Zillioux EJ, Johnson IC, Kiparissis Y, Metcalfe CD, Wheat JV, Ward SG, and Liu H. 2001. The sheepshead minnow as an *in vivo* model for endocrine disruption in marine teleosts: A partial life-cycle test with 17-ethynylestradiol. *Environ. Toxicol. Chem.* 20: 1968–78.

INDEX